Cambridge IGCSE™

Physics

Practical Skills

Heather Kennett

HODDER EDUCATION
AN HACHETTE UK COMPANY

The Publishers would like to thank the following for permission to reproduce copyright material.

Cambridge International copyright material in this publication is reproduced under licence and remains the intellectual property of Cambridge Assessment International Education. Reproduced by permission of Cambridge Assessment International Education.

Photo credits

p.6 © sciencephotos/Alamy; **p.9** © Ian Poole/iStockphoto.com

Every effort has been made to trace all copyright holders, but if any have been inadvertently overlooked, the Publishers will be pleased to make the necessary arrangements at the first opportunity.

Orders: please contact Hachette UK Distribution, Hely Hutchinson Centre, Milton Road, Didcot, Oxfordshire, OX11 7HH. Telephone: +44 (0)1235 827827. Email education@hachette.co.uk. Lines are open from 9 a.m. to 5 p.m., Monday to Friday. You can also order through our website: www.hoddereducation.com.

ISBN: 978 1 3983 1055 1

First published in 2021 by
Hodder Education,
An Hachette UK Company
Carmelite House
50 Victoria Embankment
London EC4Y 0DZ

www.hoddereducation.com

Impression number 10 9 8 7 6 5 4 3 2 1

Year 2025 2024 2023 2022 2021

Cover photo © Zffoto – stock.adobe.com

Illustrations by Integra Software Services and Aptara, Inc.

Typeset in India by Aptara, Inc.

Printed and bound by CPI Group (UK) Ltd, Croydon, CR0 4YY

A catalogue record for this title is available from the British Library.

Contents

How to use this book 4

Experimental skills and abilities 5

1 Motion, forces and energy 18
- **1.1** Simple pendulum 18
- **1.2** Density 22
- **1.3** Motion 26
- **1.4** Stretching a spring 29
- **1.5** Balancing a ruler 33
- **1.6** Centre of gravity 37
- **1.7** Pressure 40

2 Thermal physics 43
- **2.1** Supplement: Specific heat capacity 43
- **2.2** Cooling curves 47
- **2.3** Conduction and radiation 51

3 Waves 56
- **3.1** Ripple tank experiments (Teacher demonstration) 56
- **3.2** Reflection in a plane mirror 61
- **3.3** Refraction of light 66
- **3.4** Images formed by a converging lens 71
- **3.5** Speed of sound in air 76
- **3.6** Investigating pitch and loudness of sound waves (Teacher demonstration) 80

4 Electricity and magnetism 85
- **4.1** Magnetism 85
- **4.2** Electric charges (Teacher demonstration) 88
- **4.3** Measuring current 92
- **4.4** Measuring resistance 95
- **4.5** Supplement: Potential divider 100
- **4.6** Supplement: Measuring electrical power 104
- **4.7** Electromagnetism 108
- **4.8** Supplement: Investigating beams of electrons (Teacher demonstration) 113

5 Space physics 116
- **5.1** Phases of the Moon 116

Past paper questions 119
- Practical Test past paper questions
- Alternative to Practical past paper questions

Note that there are no practicals provided in this book for the Nuclear physics topic, due to safety considerations in this area. Due to this omission, we have renumbered the sections in this book, so they will differ from the section numbers in the accompanying Student's Book and Workbook.

How to use this book

This *Practical Skills Workbook* will help you to keep a record of the practicals you have completed, as well as your results and conclusions. It provides additional practice for the practical skills required by the Cambridge IGCSE™ Physics syllabuses (0625/0972) for examination from 2023.

This resource covers Core and Supplement content. Supplement investigations and questions are indicated by a lined box around the text, as shown below.

1 The average value for the refractive index of the glass block is ....

Some practicals also have Going Further sections, which provide additional questions that apply the scientific theory learned from the practical to different contexts. These go beyond Core and Supplement level, and the requirements of the syllabus, and can be used as stretch activities.

GOING FURTHER

2 Suggest, in terms of its specific heat capacity, why water is used in the radiators of central heating systems.

....

....

Completing the investigations

At the start of each investigation, we have provided a brief piece of context to help explain how the science behind the practical ties into the wider syllabus. Key terms and equations that are relevant to each investigation are also provided.

The aim of each practical is then given, along with a list of apparatus needed to complete the practical as suggested. Your teacher will inform you whether they have decided to change any of the equipment and if the method needs to be adapted as a result.

Before you begin the practical and start on the method, it is vital that you read and understand the safety guidance, as well as taking any necessary precautions. Once you have carried out a risk assessment and made everything safe, you should check with your teacher that it is appropriate to begin working through the method.

The method is presented as step-by-step instructions and you should read it through at least once before starting, making sure you understand everything. Then, ensuring that you don't miss anything out, you should work through the practical safely. Tips may be provided to help with particularly problematic steps.

Questions and answers

Within each practical, there are clear sections laid out for observations where you should record your results as you complete the practical. Scaffolded questions are also provided to help you develop conclusions and evaluate the success of the experiment.

At the end of the book are past paper questions that relate to the practicals within this book and provide useful exam practice. Your teacher may decide to set this as part of the lesson, or at a later date.

Answers to all of the questions are provided in the accompanying *Cambridge IGCSE Physics Teacher's Guide with Boost Subscription*.

Experimental skills and abilities

Skills for scientific enquiry

The aim of this book is to help you develop the skills and abilities needed to perform practical laboratory work. This section begins with an introduction to some common apparatus and measuring techniques. Some methods for making and recording accurate measurements will follow, along with descriptions of how to handle the observations and data you have collected. At the end of the section, we discuss how to plan, carry out and evaluate an investigation. You should then be ready to work successfully through the experiments and laboratory activities in this book.

Safety

Here are a few simple precautions to help ensure your safety when carrying out physics experiments in the laboratory.

- **Always wear shoes**: this will mean that your feet are protected if a heavy weight falls on them.
- **Turn off the power**: when connecting electrical circuits, ensure the power is turned off. When you are ready to take measurements, check the circuit and set the power to a low output before turning it on. Large currents can cause burns and electric shocks, and damage sensitive meters. Switch the power off between readings.
- Set up circuits away from water (taps or sink) and do not touch electrical equipment with wet hands.
- **Take care with hot liquids and solids**: set them in a safe position where they will not be accidentally knocked over; handle them with caution to avoid burns.
- **Protect eyes**: avoid looking directly into a laser beam or ultraviolet lamp; radiation can damage your eyes so wear eye protection when instructed.
- **Take care with toxic materials**: materials such as mercury are toxic; take care not to allow a mercury thermometer to roll onto the floor and break.
- **Tie back long hair** to prevent it being caught in a flame, for example.
- **Put away personal belongings**: leave them in a sensible place so that no one will trip over them!

Special note to teachers

We believe that the experiments can be carried out safely in school laboratories. However, it is the responsibility of the teacher to make the final decision depending on the circumstances at the time. Teachers must ensure that they follow the safety guidelines set down by their employers, and a risk assessment must be completed for any experiment that is carried out. Teachers should draw students' attention to the hazards involved in the particular exercise to be performed. The hazards are shown within the 'Safety guidance' section of the individual practicals.

It is recognised that, in some instances, there may not be sufficient apparatus to carry out a class practical and in such cases, the experiment could be carried out as a demonstration. (Consider using some student assistance.)

In some cases, certain pieces of apparatus may not be available. If possible, use alternatives, **as long as the safety precautions are not overlooked**. For example, a stopwatch could be used in place of an automatic timer. If you substitute equipment, consider the accuracy of the alternative; it is usually much lower than the equipment suggested.

Using and organising apparatus and materials

In an experiment, you will first have to decide on the measurements to be made and then collect the apparatus required. The quantities you will need to measure most often in laboratory work are mass, length and time.

- What apparatus should you use to measure each of these?
- Which measuring device is most suitable for the measurements you will take?
- How do you use the device correctly? For example, does the instrument need to be set to a particular range to avoid damage?

Balances

A **balance** is used to measure the mass of an object. There are several types available.

- In a beam balance, the unknown mass is placed in one pan and balanced against known masses in the other pan.
- In a lever balance, a system of levers acts against the mass when it is placed in the pan.
- A digital top-pan balance, which gives a direct reading of the mass placed on the pan, is shown in Figure 1.

Figure 1 A digital top-pan balance

The unit of mass is the **kilogram** (kg). The gram (g) is one-thousandth of a kilogram:

$$1\,\mathrm{g} = \frac{1}{1000\,\mathrm{kg}} = 10^{-3}\,\mathrm{kg} = 0.001\,\mathrm{kg}$$

How precisely do your scales measure?

- A beam balance is precise to the size of the smallest mass that tilts the balanced beam.
- A digital top-pan balance is precise to the size of the smallest mass which can be measured on the scale setting you are using, probably 1 g or 0.1 g.

Rulers

Rulers are used to measure lengths. The unit of length is the **metre** (m). Multiples are:

- 1 decimetre (dm) = 10^{-1} m
- 1 centimetre (cm) = 10^{-2} m
- 1 millimetre (mm) = 10^{-3} m
- 1 micrometre (µm) = 10^{-6} m

A **ruler** is often used to measure lengths in the centimetre range. The correct way to measure with a ruler is shown in Figure 2, with the ruler placed as close to the object as possible.

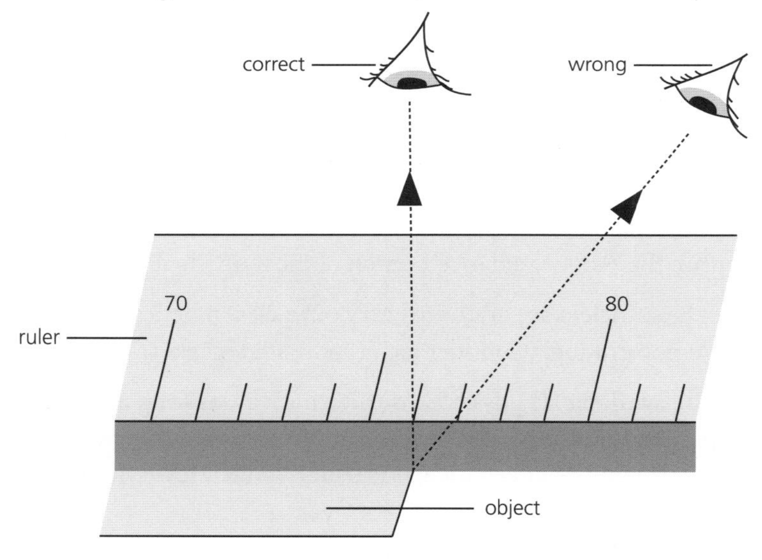

Figure 2 Using a ruler: the reading is 76 mm or 7.6 cm. Your eye must be directly above the mark on the scale or the thickness of the ruler causes parallax errors.

When measuring extensions (of springs, for example), it is best to record the actual scale readings for the stretched and the unstretched lengths, and then work out the extension afterwards.

How precise are your length measurements?

It may be possible on some rulers to estimate a measurement to the nearest half-division on the scale (0.5 mm).

For very small distances, multiples can be measured and then divided to find an average (mean) value. For example, to obtain the average thickness of one page of a book, measure the thickness of 20 pages and divide your result by 20.

Clocks and timers

Clocks, watches and timers can be used to measure time intervals. In an experiment, it is important to choose the correct timing device for the required measurement.

The unit of time is the **second** (s). A **stopwatch** will be sufficient if a time in minutes or seconds is to be measured, but if times of less than a second are to be determined, reaction times will influence the measurements.

How precise are your timings?

When using a stopwatch, reaction times will influence the reading; an accuracy of about 0.5 s is the best that is likely to be achieved. For time intervals of the order of seconds, a more precise result will be obtained by measuring longer time intervals and then dividing to find an average (mean) value. For example, to find an average value for the period of oscillation of a pendulum, time ten oscillations rather than one and then divide by 10.

To measure very short time intervals, use an automatic timer that can be triggered to start and stop by an electronic signal from a microphone, photogate or mechanical switch.

Changing measurements

If values are changing rapidly, take readings more frequently. It will often be helpful to work with a partner who watches the timer and calls out when to take a reading.

Pressing the lap-timer facility on the stopwatch at the moment you take a reading freezes the time display for a few seconds and will enable you to record a more precise time measurement.

For rapidly changing measurements of an object's motion, it may be necessary to use a **tickertape timer** (see Experiment 1.3) or a **data logger** and computer.

Some other measuring devices

Measuring cylinders

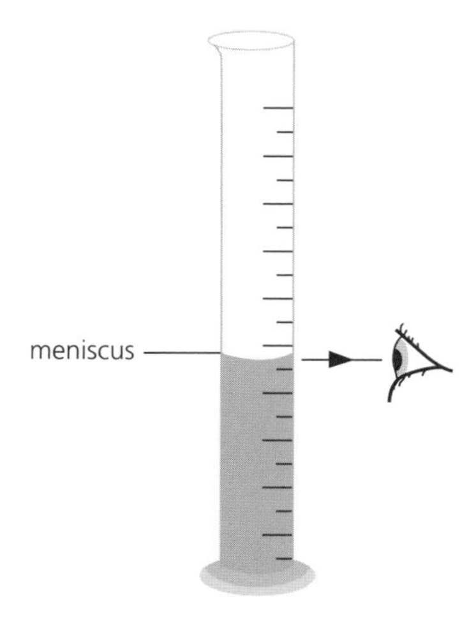

Figure 3 When taking a reading, the measuring cylinder should be vertical and your eye should be level with the bottom of the curved liquid surface – the meniscus. (For mercury, the meniscus is curved upwards; you should read the level of the top of the meniscus in a mercury thermometer.)

The volume of a liquid can be obtained by pouring it into a **measuring cylinder**. The usual units for volume are **cubic metres** (m^3), cubic decimetres (dm^3) or cubic centimetres (cm^3). Measuring cylinders are often marked in millilitres (ml) where 1 millilitre = 1 cm^3; note that 1000 cm^3 = 1 dm^3 (= 1 litre).

The precision of the reading will depend on the size of the measuring cylinder and the spacing of the scale marks. It may be possible to read to half a scale division if there is sufficient spacing in the marks of the scale.

Set squares

A **set square** is useful to determine a line at right angles to a base line. For example, in Experiment 1.4 it can be used to check whether the ruler is at 90° to a horizontal bench. It can also be used to minimise parallax errors by ensuring a reading is taken directly opposite a scale marking on a ruler.

The diameter of a cylinder can be found by setting it between two set squares aligned against a ruler (see Figure 4). A set square can also be useful for drawing parallel lines.

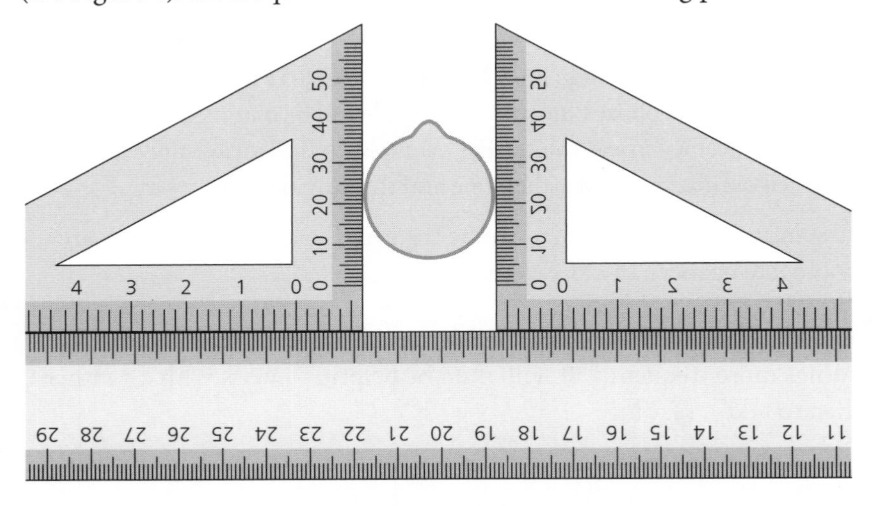

Figure 4 Using two set squares to measure the diameter of a cylinder

Protractors

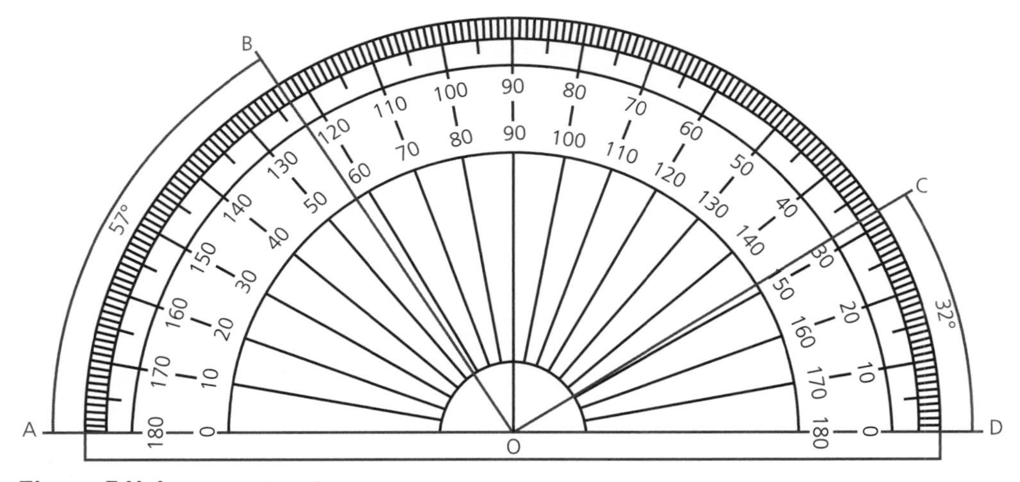

Figure 5 Using a protractor

A **protractor** is used to measure angles in **degrees** (°).

When the angle between two intersecting lines OA and OB is required, set the horizontal (0–180°) line of the protractor on one of the lines (OA) and the zero dot of the protractor on the intersection point of the lines (O). The angle AOB = 57 ° can then be read off the inner scale of the protractor as shown in Figure 5. Similarly, the angle between lines OD and OC, angle DOC = 32°, can be read from the outer scale.

The precision of the reading will depend on the size of the protractor; it is about 1° for a protractor from a school geometry set.

Ammeters and voltmeters

An **ammeter** measures electric current. The unit of current is the **ampere** (A). An ammeter should be placed **in series** with the device in which the current is to be measured and should have a low resistance so that it does not change the current to be measured.

A **voltmeter** measures potential difference (p.d.). The unit of p.d. is the **volt** (V). A voltmeter should be placed **in parallel** with the device across which the p.d. is to be measured and should have a high resistance so that it does not change the current and hence the p.d. to be measured.

Reading an analogue meter

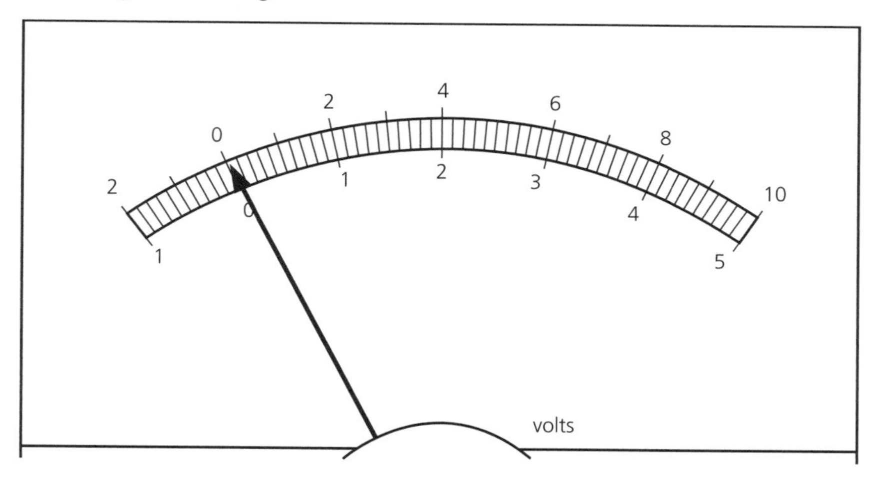

Figure 6 The display of an analogue voltmeter

Figure 6 shows the display of an analogue meter with two scales. The 0–5 scale has a full-scale deflection of 5.0 V; each small-scale division on this scale represents 0.1 V. A measurement may be able to be made to about half a small division (here, 0.05 V). For the 0–10 scale, each small division represents 0.2 V.

As with rulers, the eye should be immediately above the pointer when taking a reading to avoid introducing parallax errors; if there is a mirror behind the pointer, the needle and its image should coincide when you take a reading.

Check that the meter reads zero when there is no current; adjust the screw at the base of the pointer until it does.

Reading a digital meter

Digital meters, such as that shown in Figure 7, allow different ranges to be selected and the display gives the measurement in whatever units have been chosen. The reading will be precise to the last figure on the display, so for small currents and voltages higher precision will be achieved by using the mA or mV setting.

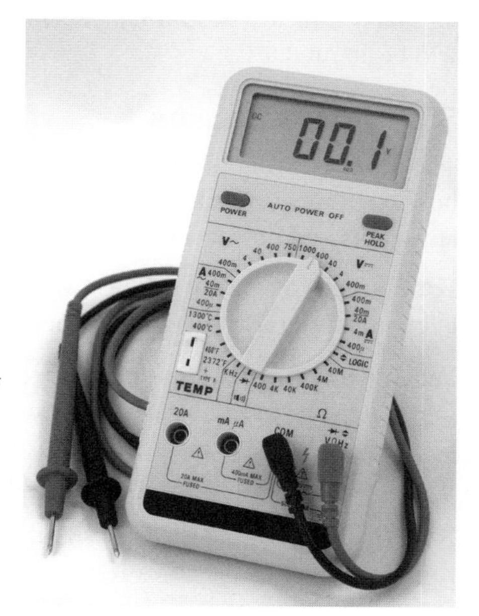

Figure 7 A digital meter

Tips for using meters in electrical circuits

- Construct circuits with the power switched off or battery disconnected and attach the voltmeter last.
- Check that meters are connected with the correct polarity and are set to their largest range initially.
- Set the power supply output to zero before you switch it on.

Observing, measuring and recording

Having collected together and familiarised yourself with the equipment and materials needed for an experiment, you are now ready to start making some observations and measurements.

- Decide on the range and interval of readings you will take.
- If the method does not include a diagram of the apparatus, it may be helpful at this stage to draw a clearly labelled diagram of the experimental set-up.
- You should also record any difficulties encountered in carrying out the experiment and any precautions taken to achieve accuracy in your measurements.
- Do not dismantle the equipment until you have completed the analysis of your results and are sure you will not have to repeat any measurements!
- How precise will your measurements be?
- How many significant figures will your data have?
- How will you record your results?

Precision of measuring devices

Make a list of the apparatus you use in an experiment and record the smallest division of the scale of each measuring device; since you will usually be able to take a measurement to half a small division, this will be the **precision** of your measurements.

For example, the smallest division on a metre ruler is 1 mm, so the precision of any length measured with the ruler will be about half a division (0.5 mm). The precision will be a smaller proportion of a measurement the longer the length measured:

- For a measured length of 1 m = 1000 mm, the precision will be 1 part in 2000.
- For a measured length of 1 cm = 10 mm, the precision will be 1 part in 20.

Similarly if the divisions on a thermometer scale are at 1°C intervals, the precision of a temperature reading will be about 0.5°C.

Significant figures

The number of digits given for a measurement or calculated value, called **significant figures**, indicates how accurate we think it is. You should not give more digits in a calculated answer than are justified by the apparatus and how it was used.

For example, a value of 6.7 has two significant figures; the value of 0.235 has three significant figures, the 2 being most significant and the 5 being the least significant.

When doing calculations your answer should have the same number of significant figures as the measurements used in the calculation. For example, if your calculator gives an answer of 1.23578, this should be given as 1.2 if the measurements on which you based this calculation have two significant figures and 1.24 if your measurements have three significant figures.

Note that in deciding the least significant figure you look at the following digit; if that is less than 5, you round down (1.23 becomes 1.2) but if it is 5 or above, you round up (1.235 becomes 1.24).

If a number is expressed in standard notation, the number of significant figures is the number of digits before the power of 10; for example, 6.24×10^2 has three significant figures.

If values with different numbers of significant figures are used to calculate a quantity, quote your answer to the smallest number of significant figures.

Sources of error

Every measurement of a quantity is an attempt to find its true value and is subject to errors arising from the limitations of the apparatus and the experimental procedure.

Random errors

Limitations on the precision of a measuring device may produce random errors. If readings fall between scale markings and have to be rounded up or down, a scatter of results occurs. Experimental difficulties such as variable reaction times or fluctuating environmental conditions may also produce random errors. Repeating measurements and taking the mean (see page 12) will help to average out random errors and reduce the uncertainty in a measurement.

Systematic errors

Figure 8 shows part of a ruler used to measure the height of a point, P, above the bench.

- The ruler has a space of length x before the zero of the scale.
- The height of the point P = scale reading + x = 5.9 + x.
- By itself the scale reading is not equal to the height of P; it is too small by the amount x.

An error of this type is called a **systematic error** because it is introduced by the system used to make the measurement. A half-metre ruler does not generally have a systematic error because its zero is usually at the end of the ruler. When using a ruler to measure a height, the ruler must be held so that it is vertical. If it is at an angle to the vertical, a systematic error will be introduced.

Before making a measurement, check to ensure that the reading is zero, otherwise a zero error must be allowed for when a reading is taken. This can often happen with a top-pan balance or a stopwatch, for example. See also the section on ammeters and voltmeters (page 9).

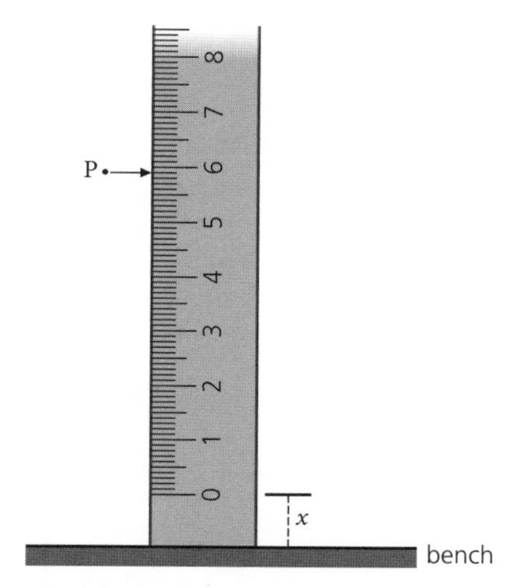

Figure 8 Introducing a systematic error into a measurement

Tables

If several measurements of a quantity are being made, draw a table in which to record your results.

- Use the column headings, or start of rows, to name the measurement and state its unit. For example, in Experiment 1.1 (see page 18) you will use a table similar to Table 1 to record your results.
- Repeat the measurement of each observation if possible and record the values in your table; if repeat measurements for the same quantity are significantly different, take a third reading. Calculate an average value from your readings. If you decide not to include an apparently anomalous value when calculating your mean, state that it has been omitted and suggest a reason for its occurrence.
- Numerical values should be given to the number of significant figures appropriate to the measuring device.

Table 1

Mass of pendulum bob/g	First measurement of pendulum length, L_1/cm	Second measurement of pendulum length, L_2/cm	Average pendulum length, L/cm

Handling experimental observations and data

Once you have your measurements, you will need to process them. You may need to make calculations or plot a graph of your results. Then you can summarise what have you learnt from the experiment, discuss sources of experimental error and draw some conclusions from the investigation.

- What is the best way to process your results?
- Are there some anomalous values to be dealt with?
- What experimental errors are there?
- What conclusions or generalisations can you make? What patterns can you see?

Calculations

You may have to produce an average value or manipulate an equation to process your results.

Averages (mean)

Sum the values for a quantity you have measured and divide the sum by the number of values to obtain the average. For example, if you measure the length of a pendulum as 81.5 cm and 81.6 cm, then the mean value is:

$$\frac{(81.5 + 81.6)}{2}\,\text{cm} = \frac{163.1}{2}\,\text{cm} = 81.55\,\text{cm} = 81.6\,\text{cm}$$

The value has been given to three significant figures because that was the precision of the individual measurements on which the calculation was based.

Equations

When tackling physics problems using mathematical equations, do not substitute numerical values until you have obtained the expression that gives the answer in symbols. This reduces the chances of making arithmetic and copying errors.

Equations frequently have to be rearranged to change the subject. For example, in the equation $F = kx$, the subject is F. To change the subject to x, we must divide both sides by k so that:

$$\frac{F}{k} = \frac{kx}{k} = x$$

or

$$x = \frac{F}{k}$$

If you have numerical values for F and k, you can now substitute them into the equation to calculate x. For example, if $F = 4.16\,\text{N}$ and $k = 2.0\,\text{N/cm}$:

$$x = \frac{F}{k} = \frac{4.16}{2.0} = 2.08\,\text{cm} = 2.1\,\text{cm}$$

The value for x is given to two significant figures because that was the lower of the number of significant figures for the values of F and k on which the calculation was based.

Graphs

Graphs can be useful in finding the relationship between two quantities.

- You will need about six data points taken over as large a range as possible to plot a graph.
- Choose scales that make it easy to plot the points and use as much of the graph paper as possible.
- Plot the dependent variable along the y-axis and the independent variable along the x-axis, unless directed otherwise.
- Make sure you label each axis of the graph with the name and unit of the quantity being plotted.
- Mark the data points clearly with a dot in a circle (⊙), a plus sign (+) or a cross (×) with a sharp pencil.
- Join your points with a smooth line or curve of best fit.

Straight-line graphs

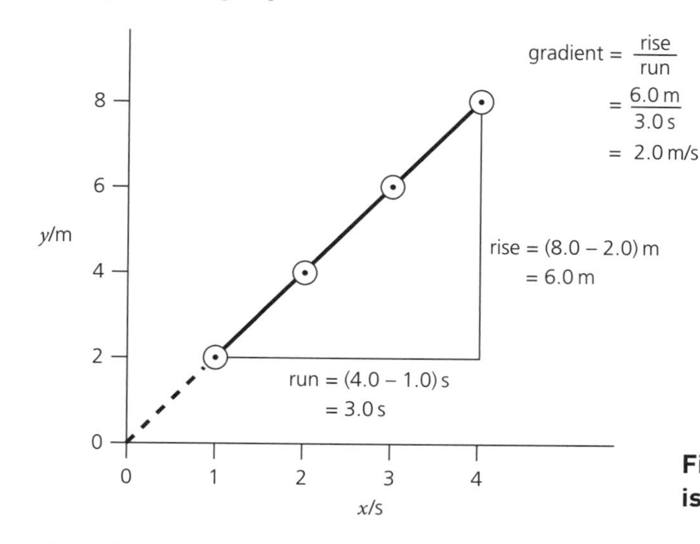

x/s	y/m
1.0	2.0
2.0	4.0
3.0	6.0
4.0	8.0

Figure 9 A graph showing that y is directly proportional to x

When the readings in the table above are used to plot a graph of y against x, the line of best fit joining the points (Figure 9) is a straight line through the origin (0, 0). Such a graph shows that there is direct proportionality between the quantities plotted: $y \propto x$. Note, however, that the line must go through the origin for the quantities to be proportional.

If a straight-line graph does not go through the origin, one can only say that there is a linear relationship between y and x. In such a case, the value of y when it crosses the y-axis is termed the y-intercept; the value of x at the y-intercept is always zero.

The equation of a straight-line graph can be written in the form

$$y = mx + c$$

where m is the gradient and c is the value of the y-intercept.

The x-intercept is the value of x where the line crosses the x-axis; y is always zero at that point.

Slope or gradient

The slope or gradient of a straight-line graph can be determined by the triangle method shown in Figure 9. Use as long a length of line as possible to determine the gradient from the ratio of the vertical 'rise' to the horizontal 'run' of the triangle chosen. The units of the gradient are determined by the units of the quantities plotted on the graph axes.

If the graph is a curve, the gradient at a particular point on the curve is found by drawing a straight-line tangent to the curve at that point and again calculating the gradient by the triangle method.

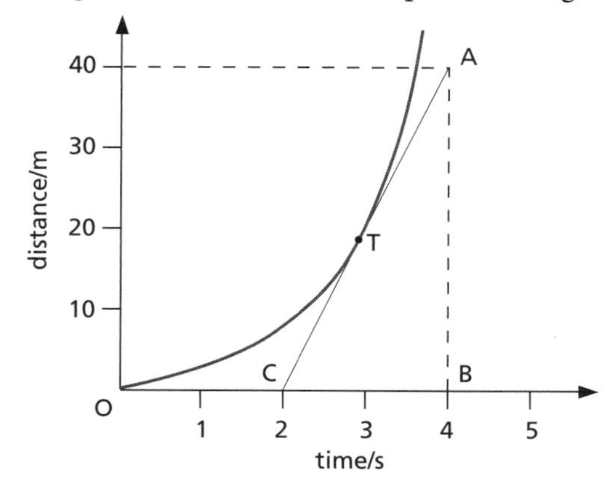

Figure 10 Tangent to a curve

In Figure 10, the gradient of the tangent to the curve at point T is:

$$\frac{AB}{CB} = \frac{40\,\text{m}}{2\,\text{s}} = 20\,\text{m/s}$$

Errors

In practice, points plotted on a graph from actual measurements may not lie exactly on a straight line due to experimental errors. In these cases, the 'best straight line' is drawn through the points such that they are equally distributed about it; this takes account of the uncertainties in the measurements. Do not force the line through the origin.

Anomalous data are readings that fall outside the normal or expected range of measurements. If possible, repeat any anomalous measurements to check that they have been recorded properly or try to identify the reason for the anomaly.

Conclusions

Once you have analysed your experimental results, summarise your conclusions clearly and relate them to the aim of the experiment. State whether a hypothesis has been verified; suggest reasons if your results do not, or only partially, support a hypothesis.

If a numerical value has been obtained, state it to the correct number of significant figures. Compare your results with known values if available and suggest reasons for any differences.

State any relationships discovered or confirmed between the variables you have investigated. Mention any patterns or trends in the data.

Planning an investigation

When planning an experiment it is important to set a specific aim and produce a logical plan for the investigation.

Identify the **control variables** in the investigation and decide which ones to investigate and which ones you should try to keep **constant** (fixed) so that they do not affect the experimental results. The variable that is changed is known as the **independent variable**. The variable that is measured is known as the **dependent variable**.

To discover the relationship between variables, you should change only one variable at a time. For example, in Experiment 1.1 (see page 18), when investigating the variation of the period of oscillation of a pendulum, you first keep the mass of the bob constant and record the period for different pendulum lengths; this will reveal how the period depends on pendulum length. You then keep the length of the pendulum constant and measure the period when different masses are used; this will show how the period depends on the mass of the bob. Other variables, such as the amplitude of the oscillation, may affect your results, so identify these before you start the experiment so that you can try to keep them constant.

Write the plan for your investigation as a series of step-by-step instructions which should include:

- the apparatus and materials you will need; make sure your measuring devices have enough precision for the measurements you will make
- the variable you will change
- the variable you will measure
- the variables you will keep the same
- the number of readings or measurements you will take
- the range of values of the independent variable and the time allowed for any readings to take place.

Before you start the experiment, familiarise yourself with how to use the apparatus. It will be helpful to decide how to record your results; draw tables in which to record your measurements if appropriate.

Evaluating investigations

As a final stage, you should evaluate the experiment and discuss how it could be improved.

Identify and comment on sources of error in the experiment. For example, it may be very difficult to eliminate all energy losses to the environment in an experiment where the temperature rise of an object is measured; if that is the case, say so. Mention any sources of systematic error in the experiment, for example a zero error on a meter or measuring device. Reaction times may also have reduced the accuracy of measurements. It may not have been possible to control some variables in the experiment, for example keeping the temperature of a current-carrying wire constant. Try to identify these types of error and suggest improvements that would eliminate or reduce the effect of these errors.

Could some things have been done better? For instance, do you have enough results to show a pattern? Was the range of the independent variable wide enough? Could you have repeated or collected further data in one part of the range where the results were uncertain or anomalous? If so, suggest changes or modifications that could be made to the procedure or to the equipment used in the investigation.

Questions

1 What measuring device would you use to obtain values for each of the following?

a the volume of liquid in a coffee mug ....

b the mass of an apple ....

c the length of an envelope ....

d the diameter of a tin of tomatoes ....

e the temperature of a cup of tea ....

f the time taken to run up 20 stairs ....

g the p.d. across a lamp ....

h the time taken by an apple to fall through one metre ....

i the angle that a ray of light is turned through by a plane mirror ....

j the current flowing in a resistor ....

k the dimensions of a textbook ....

2 How would you obtain a value for each of the following?

a the average thickness of a newspaper page

....

....

b the average time for one oscillation of a child's swing

....

....

c the average mass of a pin

....

....

3 Complete the table below by stating the typical precision of each of the measuring devices listed.

Device	metre ruler	stopwatch	digital timer	top-pan balance	liquid-in-glass thermometer	100 cm^3 measuring cylinder
Precision						

4 Write the number 9.753864 to:

a three significant figures ....

b two significant figures ....

c one significant figure ....

5 The power of a device is calculated from the equation:

$P = IV$

a If $I = 250 \times 10^{-3}$ A and $V = 8.0$ V, calculate P, giving your answer to the correct number of significant figures.

b If $P = 60$ J/s and $V = 12$ V, calculate I.

6 The reading on an ammeter when no current is flowing in it is 2 mA. What is the true value of the current if the meter reads 26 mA when a current flows in it?

7 The measurements in the table below were obtained for the speed of a trolley rolling down a runway from rest.

Speed/m/s	0.0	0.2	0.4	0.6	0.8	1.0
Time/s	0.00	0.05	0.10	0.15	0.20	0.25

a i State the variables being measured.

....

ii Name a variable that is being kept constant.

....

iii Complete the table below.

Independent variable (variable you change)	**Control variables (constant)**	**Dependent variable (variable you measure)**

b Plot a graph of time on the horizontal axis and speed on the vertical axis.

c Calculate the gradient of the graph.

d What can you conclude about the relationship between speed and time in this experiment?

....

....

1 Motion, forces and energy

1.1 Simple pendulum

Mass, length and time are the three basic quantities we measure in physics. All three are measured in this experiment. You will often measure these quantities in your practical work and for each application you should choose the appropriate measuring device. To reduce the effect of random errors on the accuracy of a measurement, measure multiples and find the average (mean) value (see page 7). The time for one oscillation of a simple pendulum (the period) is too short to be measured accurately with a stopwatch, so the longer time for several oscillations is measured and divided by the number of oscillations to obtain a mean value for one oscillation.

KEY TERMS

Period
Dependent variable
Independent variable
Control variable
Accuracy

Aim

To investigate the effect of length and mass on the period of a simple pendulum.

Apparatus

- Support stand
- Clamping plates (wood or metal)
- String
- Two metal pendulum bobs of different mass
- Stopwatch or clock
- Metre ruler
- Balance

SAFETY GUIDANCE

This practical presents minimal risk.

Method

You will investigate the effect of length and the effect of mass on the period of a simple pendulum. Before you take any measurements, read the instructions below. For each experiment, complete Table 1 to identify which variable you will change, which variables you will keep constant and which variable you will measure.

Table 1 Determining variables

	Independent variable (variable you change)	Control variables (constant/fixed)	Dependent variable (variable you measure)
Experiment 1			
Experiment 2			

Set up the apparatus as shown in Figure 1.

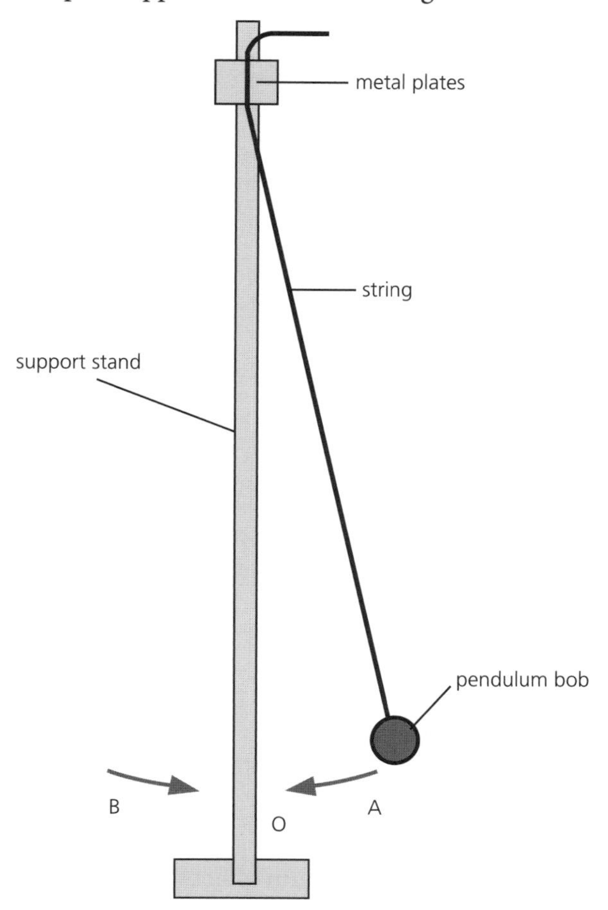

Figure 1

1 Measure the length of the pendulum from the point of support to the centre of the bob; repeat your measurement and calculate the average length. Record your results in Table 2.

2 Pull the pendulum bob slightly to one side and let it swing. Measure the time for the bob to make ten oscillations; the amplitude of the oscillations should be small.

One complete oscillation occurs when the bob moves from A → O → B → O → A. Repeat the measurement and calculate the average time for ten oscillations. Record your results in Table 3.

3 Calculate the period, *T*, of the pendulum; this is the time needed for one oscillation.

4 Measure the mass of the pendulum bob.

5 Repeat steps **1** to **4** using a longer pendulum length (first experiment).

6 Repeat steps **2** to **4** using a heavier pendulum bob and the same pendulum length as was used in step **5** (second experiment).

TIP 

Keep the angle of release the same each time so that the amplitude of the oscillations remains constant.

TIP 

It may be helpful to place a marker behind the pendulum to ensure the ball is released from the same point each time and small amplitude oscillations are used.

Observations

Record your results in Tables 2 and 3.

Table 2

Mass of pendulum bob/g	First measurement of pendulum length/cm	Second measurement of pendulum length/cm	Average pendulum length/cm

Table 3

Mass of bob/g	Average pendulum length/cm	First measurement of time for ten oscillations/s	Second measurement of time for ten oscillations/s	Average time for ten oscillations/s	Period, *T*/s

Conclusions

What do your results suggest about how the length and the mass of a simple pendulum affect the period of the pendulum?

1 Length: ....

....

2 Mass: ....

....

Evaluation

1 Do you think you had enough results to justify your conclusions? Explain your answer.

....

....

....

2 Explain why ten oscillations are timed rather than one oscillation.

....

....

....

3 Suggest how your measurement of the length of the pendulum could be improved.

....

....

....

GOING FURTHER

1 The force of gravity causes the pendulum bob to fall towards the centre of the oscillation (O). A value for the acceleration of free fall, g, can be calculated from the period of a simple pendulum using the following equation:

$$g = \frac{4\pi^2 l}{T^2}$$

where l is the length of the pendulum in metres, measured from the point of suspension to the centre of mass of the bob, and T is the period, in seconds.
Calculate g using one set of your results.

g = ... m/s^2

2 Describe how you could verify the relationship between T^2 and l graphically.

....

....

....

....

1.2 Density

The density of a substance is defined as the mass per unit volume and is given the unit g/cm^3.

Different materials have different densities; those with a lower density than water will float in water while those with a higher density will sink.

To determine the density of a material, both its mass and volume must be known. For a regularly shaped solid, the volume can be found by measuring its dimensions. For an irregularly shaped solid which sinks in water, the volume is found by the displacement method. The solid displaces the same volume of water as its own volume.

Aim

To measure the density of some liquids and solids.

Apparatus

- Eye protection
- Measuring cylinder
- Ruler
- Balance (minimum resolution 1 g)
- Solid, regularly shaped blocks of different materials (A and B), Cylinder (C), Ball (D)
- Irregularly shaped solid (E), which will fit inside the measuring cylinder
- Water

Method

Regularly shaped solids

1. Select blocks A and B, which are made of two different materials.
2. Measure the dimensions of each block and record the values in Table 1 in the Observations section. Include the units of each measurement you make.
3. Measure and record the mass of each block.
4. Select the cylinder C; measure and record its mass and dimensions in Table 2.
5. Select the ball D; measure and record its mass and diameter in Table 2.

Liquid

1. Measure the mass of an empty measuring cylinder and record the value in Table 3.
2. Pour some water into the measuring cylinder and place it on the balance again.
3. Record the new mass of the measuring cylinder plus water.

KEY TERMS 

Density
Meniscus
Precision
Significant figures

KEY EQUATIONS 

$$\rho = \frac{m}{V}$$

where ρ = density, m = mass and V = volume.

volume of a cylinder = length × $\frac{\pi d^2}{4}$
where d = diameter of cylinder.

SAFETY GUIDANCE 

Eye protection must be worn.

TIP 

To measure the diameter of a ball, work in pairs. One person places the ball between two smooth vertical surfaces and the other measures the distance between the surfaces with a ruler.

4 Find the mass of the water. This is the difference between the first and second readings.

5 Measure and record the volume of water in the cylinder.

Irregularly shaped solid

1 Measure and record the mass of object E in Table 4.

2 Pour some water into a measuring cylinder and record its volume.

3 Place object E in the measuring cylinder so that it is covered with water.

4 Record the volume of water + E.

5 Find the volume of E. This is the difference between the first and second readings.

TIP

Your eye must be level with the bottom of the meniscus when reading the volume of water in a measuring cylinder.

Observations

Complete Tables 1, 2, 3 and 4 using the data from your experiment.

TIPS

Values should be recorded to at least two significant figures. (see page 10).

Remember to include the units in the column headings.

Regularly shaped solids

Complete the following equations.

1 Volume of a block = length × ....

2 Volume of a cylinder = length × ....

To find the volume of the ball D, you will need to use the equation:

volume of sphere $= \frac{4}{3}\pi r^3$

where r = radius of the sphere.

Table 1 Finding the density of regularly shaped blocks

Object	Length/	Width/	Height/	Volume/	Mass/	Density/
A						
B						

Table 2 Finding the density of a cylinder and a ball

Object	Length/	First measurement of diameter/	Second measurement of diameter/	Average diameter/	Volume/	Mass/	Density/
C							
D	n/a						

Liquid

Table 3 Finding the density of water

Mass of empty cylinder/	Mass of cylinder + water/	Mass of water/	Volume of water/	Density of water/

Irregularly shaped solid

Table 4 Finding the density of an irregularly shaped solid

Mass of E/	Volume of water/	Volume of water + E/	Volume of E/	Density of E/

Conclusions

1 Complete Table 5. Include the units in the column heading for density. The density of the material should be given to two significant figures.

Use the density data in your textbook to identify possible materials from which A, B, C, D and E are made.

Table 5

Object	Density/	Material
A		
B		
C		
D		
E		
Water		Water

2 Describe how you could determine whether a teaspoon is made of silver or steel.

....

....

....

....

....

....

Evaluation

1 Fill in the missing words in the following sentences.

The dimensions of the blocks A and B were measured with a ruler to a precision of ....

The diameters of the cylinder C and ball D were measured with a ruler to a precision of ....

The volume of object E was measured in a measuring cylinder to a precision of ....

Mass was measured with a balance to a precision of ....

2 Discuss how the precision of your measurements could be improved.

....

....

....

GOING FURTHER

Hang a metal mass on a newton meter and measure its weight in air. Half fill a measuring cylinder with water and record the volume of water. Lower the mass into the water with it still attached to the newton meter and take the reading again; make sure that the mass does not touch the sides or bottom of the measuring cylinder. Record the new water level in the measuring cylinder and calculate the volume and weight of water displaced by the mass.

(Take the density of water to be 1.0 g/cm^3.)

Newton meter reading when mass in air = ... N

Newton meter reading when mass submerged in water = ... N

Change in reading = ... N

Upward force on mass in water = ... N

Initial measuring cylinder reading = ... cm^3

Final measuring cylinder reading = ... cm^3

Volume of water displaced = ... cm^3

Weight of displaced water = ... N

Compare the value of the upward force on the mass when it is submerged in the water with the weight of water it displaced and comment on your result.

....

....

....

....

....

....

1.3 Motion

Average speed is defined as the ratio of the total distance travelled to the total time taken. If speed increases, an object is accelerating.

In experiments to study speed and acceleration you need to measure time and distance over short intervals of time. You can do this by using a tickertape timer.

The tickertape timer makes dots on a paper tape at specific time intervals. When the tape is attached to a moving object, the tape provides a convenient method of recording both short time intervals and distance at the same instant. Tape charts constructed by sticking consecutive tape sections vertically side-by-side show how the speed changes.

Aims

To measure the speed of a trolley at different points on a slope and to identify when an object is accelerating.

To calculate the acceleration.

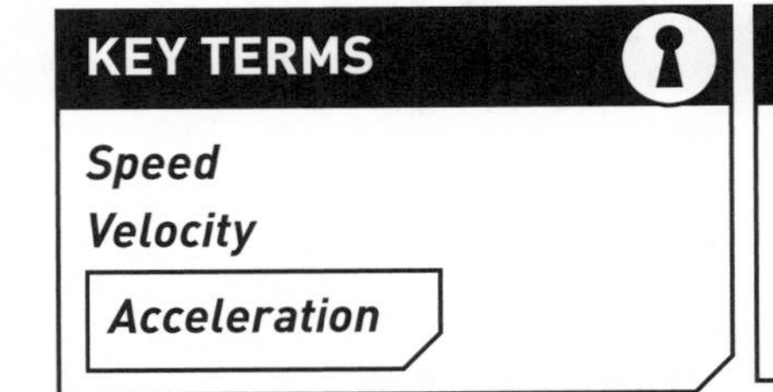

SAFETY GUIDANCE

Ensure that the trolley does not fall from the end of the runway, as it could cause injury to legs or feet.

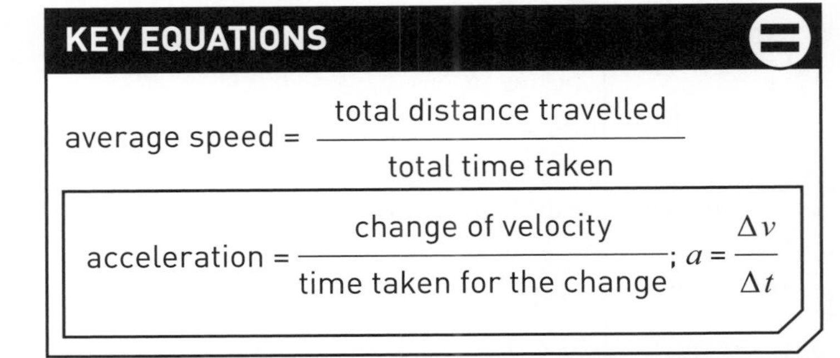

Apparatus

- Tickertape timer with power supply
- Tickertape
- Trolley
- Runway
- Ruler
- Glue or sticky tape

Method

Speed

1 Familiarise yourself with the use of the tickertape timer.

The distance between two ticks on the tape is the distance the trolley has moved in $\frac{1}{50}$ s and can be measured with a ruler.

2 With the timer switched on, pull a 1 m length of tape through the timer, first slowly at constant speed and then faster at increasing speed.

3 Cut the tape into ten-tick lengths and make a tape chart by arranging successive strips of tape side by side, so that the bottom of each strip is on a horizontal line as shown in Figure 1.

TIP

The tickertape timer produces a dot or 'tick' every $\frac{1}{50}$ s.

TIP

Cut the tape on the dots every 11th dot. This gives ten spaces between each cut.

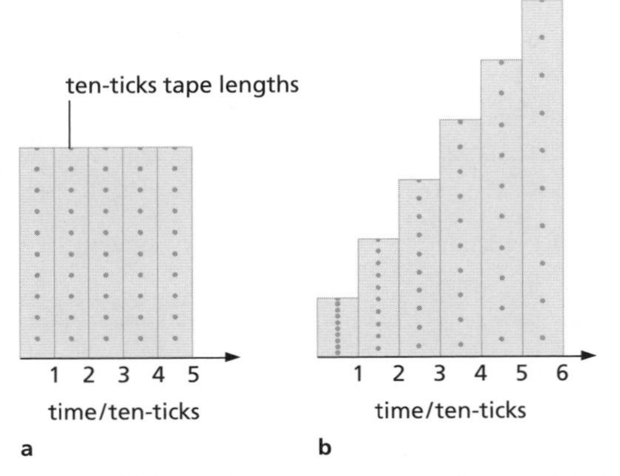

Figure 1 Tape charts: a constant speed; b increasing speed

Acceleration

1 Set up a sloping runway, tickertape timer and trolley as shown in Figure 2.

2 Attach a length of tape to the trolley and release it at the top of the runway.

3 Ignore the region at the start where the dots are very close together, but beyond them cut the tape into ten-tick lengths and make a tape chart.

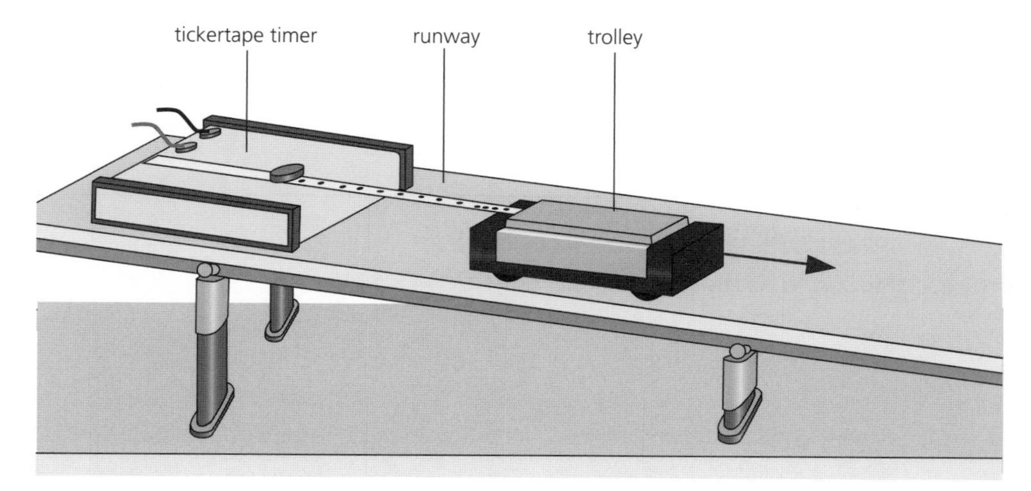

Figure 2

Observations

Speed

1 Mark the region where the tape was pulled quickly. Then mark the region where the tape was pulled slowly.

2 How can you tell when the tape is being pulled quickly?

....

....

....

....

3 Find a region of the chart where the speed was constant. Measure the distance travelled by the trolley at constant speed in one 'ten-tick' with a ruler. Use this to calculate the speed (in cm/s) of the trolley.

TIP

The time a timer takes to produce a 'ten-tick' equals:

$10 \times \frac{1}{50}\,\text{s} = \frac{1}{5}\,\text{s}.$

Speed of the trolley = ... cm/s

Acceleration

1 For two adjacent strips, measure the distance the trolley moved in 'ten-ticks' and calculate the average speed of the trolley, v_1 and v_2, in each.

....

....

....

....

2 Use the two average speeds to calculate the acceleration of the trolley in cm/s^2.

(Take the time for the acceleration to be the time for one 'ten-tick'.)

Conclusions

1 Summarise your results.

....

....

....

2 State the value you calculated for the constant speed at which you pulled the tape.

....

3 State the values you calculated for the average speed of the trolley on consecutive tapes.

v_1 = ... v_2 = ...

4 State the value you obtained for the acceleration of the trolley between two consecutive tapes.

Acceleration = ...

5 Is the acceleration of the trolley constant? Justify your answer.

....

....

....

Evaluation

1 Describe any sources of error in the experiment.

....

....

....

....

2 Describe how the experiment could be improved to give more reliable results for measuring speed.

....

....

....

1.4 Stretching a spring

A spring will stretch when a mass is hung from it. Over 300 years ago Robert Hooke discovered that provided a spring is not permanently stretched, the extension is proportional to the stretching force. Straight metal wires or elastic bands which are stretched behave in the same way. Springs of different materials and dimensions will extend by different amounts for a given stretching force.

The amount the spring extends depends on the value of its spring constant which is defined as the force per unit extension.

Aim

To plot a load–extension graph for a spring.

To determine the spring constant of a spring.

KEY TERMS 

Extension
Load
Weight
Gravitational field strength
Gradient
Significant figures
Spring constant

Apparatus

- Eye protection
- Retort stand, clamp and clamp rod
- Spring
- Hanger with five 100 g masses (or 1 N weights)
- Ruler
- Pointer (a small piece of wire)
- Adhesive/sticky tape

KEY EQUATION 

$$k = \frac{F}{x}$$

where k is the spring constant, F is the stretching force (load) and x is the extension of the spring.

Method

You will investigate the extension produced on a spring by loading it with different weights. Before you take any measurements, read the instructions below. Complete Table 1 to identify which variable you will change, which variables you will keep fixed and which variable you will measure.

Table 1

Independent variable (variable you change)	Control variables (constant/fixed)	Dependent variable (variable you measure)

SAFETY GUIDANCE 

- Eye protection must be worn.
- Take care with masses and think about where the hanging masses would fall if the spring snapped.

Set up the apparatus as shown in Figure 1.

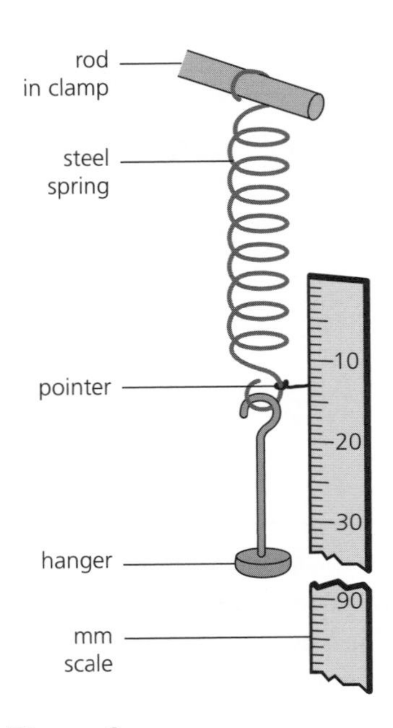

Figure 1

1. Hang the spring on a rod held in a clamp on a stand. Fix the ruler vertically next to the spring so that it can be used as a scale.
2. Attach a pointer to the lower end of the spring to help you take accurate readings of the position of the lower end of the spring.
3. Record the scale reading, l_0, of the lower end of the unweighted spring and repeat your measurement. Record the average position in the Observations section.
4. Hang an unweighted (100 g) hanger on the spring and record in Table 2 the new scale position of the lower end of the spring, as indicated by the pointer.
5. Add a 100 g mass to the hanger and again record in Table 2 the scale position of the lower end of the spring.
6. Repeat step **5** with a total of 200 g, 300 g, 400 g and 500 g masses on the hanger.

TIP 

Tape the ruler to the bench, or clamp it to the stand, so that it doesn't change position while you are taking measurements.

TIP 

Take your reading with your eye directly opposite the mark on the ruler to reduce parallax errors (see page 6).

TIP 

Record readings for the lower end of the spring and then calculate the spring extension afterwards. This will help to avoid subtraction errors.

TIP 

Add each mass slowly so that the load does not bounce up and down, and do not add more than a total of 500 g.

Observations

1. Complete the following.

   Average scale reading of lower end of unweighted spring: l_0 = ... mm

2 Complete Table 2. Note that 100 g has a weight of 0.98 N if the gravitational field strength *g* is taken as 9.8 N/kg.

Use the relation:

extension of spring = scale reading of lower end of spring − l_0

Table 2

Mass/g	Stretching force/N	Scale reading/mm	Extension/mm

3 Plot a load–extension graph for this spring. The stretching force should be along the *y*-axis with the extension in mm along the *x*-axis.

TIP 

Choose an easy to read/plot scale for your graph, mark points clearly, label axes correctly (including units) and use as much of the graph paper as possible.

4 Draw a line of best fit and calculate the gradient of the graph to an appropriate number of significant figures.

Gradient = ...

TIP

Use the triangle method (see page 13) to determine the gradient of your graph from as long a line as possible and show your working.

Conclusions

1 State whether or not the extension is directly proportional to the stretching force on the spring and justify your statement.

....

....

....

2 a Explain how the spring constant k can be determined from your graph.

....

....

....

b State the value of the spring constant, including its units.

....

....

Evaluation

1 State the precautions you took to improve the accuracy of the experiment.

....

....

....

....

....

....

2 Explain how you minimised parallax errors when reading the length of the spring on the ruler.

....

....

....

GOING FURTHER

Predict, with a reason, what results and shape of graph you would expect if you repeated the experiment using a spring of the same material but made of thinner wire.

....

....

....

....

If you have time, carry out the experiment to test your prediction.

1.5 Balancing a ruler

The moment of a force is a measure of its turning effect and is given by:

moment of a force = force × perpendicular distance from the pivot

For an object in equilibrium, the principle of moments states that the sum of the anticlockwise moments about any point equals the sum of the clockwise moments about the same point.

An object is in equilibrium when the principle of moments applies and there is no resultant force on the object.

KEY TERMS 

Equilibrium
Moment
Principle of moments
Weight

KEY EQUATION 

moment of a force = $F \times d$

where F is the turning force acting on a body and d is the perpendicular distance of the line of action of the force from the pivot.

Aim

To measure the moment of a force about a pivot and to show that there is no resultant moment on a body in equilibrium.

Apparatus

- Retort stand and clamp
- Half-metre ruler with a central drilled hole
- Long nail
- Three hangers with sets of 10 g and 20 g masses
- Balance
- Blu-tack or adhesive/mounting putty
- String or rubber band

Method

In this experiment you will balance a ruler by hanging masses at different distances from the pivot and calculate the moment of the forces about the pivot when the ruler is in equilibrium. Set up the apparatus as shown in Figure 1, with the pivot (nail) supported horizontally in a retort stand.

SAFETY GUIDANCE

Take care not to allow masses to fall off the ruler and hit the bench or floor.

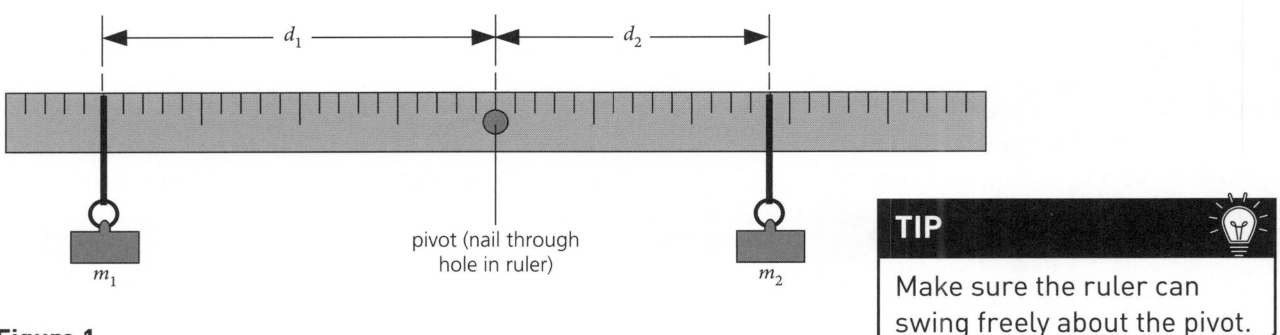

Figure 1

TIP

Make sure the ruler can swing freely about the pivot.

1. Balance the half-metre ruler at its centre, adding blu-tack to one side or the other until it is as near to being balanced as possible.
2. Hang a 30 g mass (m_1) from the ruler at a distance (d_1) of 20.0 cm from the pivot.

Figure 2

TIP

Use a loop of string (or a rubber band) to hang the mass hanger on the ruler. The string loop shown in Figure 2 allows the lower part to be pulled tight to hold the mass, and the upper part to slip over the ruler.

3. Continue to support the ruler horizontally and hang a 40 g mass (m_2) on the opposite side of the pivot. Allow the ruler to swing slightly and adjust the position of the 40 g mass so that the ruler is as balanced as possible. Record the positions of m_1 and m_2 on the ruler in Table 1. Work out the distance d_1 (distance of m_1 from the pivot) and d_2 (the distance of m_2 from the pivot) later.
4. Move m_1 5.0 cm nearer to the pivot and adjust the position of m_2 until the ruler is again balanced; record the new positions of the masses.
5. Repeat step **4**.
6. Change the masses m_1 and m_2 to 50 g and 60 g and repeat steps **2** to **5**.

TIP

Support the ruler in a horizontal position when the mass is being attached. After recording the position of a mass, use a small amount of blu-tack or tape to help hold the string or hanger in place.

7 Select masses of 30 g, 40 g and 50 g (M_1, M_2 and M_3) and balance the ruler with two masses (at different positions) on one side of the pivot and one mass on the opposite side; record the position of each of the masses in Table 2.

TIP 

Note that for a reading of 40 cm on the ruler, the distance d_2 = (40 − 25) cm = 15 cm since the ruler is balanced at the 25 cm position.

Observations

Complete the following tables. Note that 10 g has a weight of 0.1 N if *g* is taken as 10 N/kg.

Table 1

m_1/g	F_1/N	Position on ruler/cm	d_1/cm	$F_1 \times d_1$/N cm	m_2/g	F_2/N	Position on ruler/cm	d_2/cm	$F_2 \times d_2$/N cm	Resultant moment/N cm

Table 2

	Mass/g	*F*/N	Position on ruler/cm	*d*/cm	$F \times d$/N cm	Anticlockwise moment/N cm	Clockwise moment/N cm
M_1							
M_2							
M_3							

Conclusions

1 When the ruler is balanced, what do your results suggest about the values of the anticlockwise and clockwise moments?

....

....

....

2 When the ruler is balanced, what do your results suggest about the resultant moment on the ruler?

....

....

....

Evaluation

1 Explain the reason for making sure the ruler was balanced when there were no masses hanging on it.

....

....

....

2 Describe any sources of error in the experiment.

....

....

....

GOING FURTHER

Describe how you could use the apparatus to obtain the mass of an unknown object, such as an apple.

....

....

....

....

1.6 Centre of gravity

An object behaves as if its whole mass were concentrated at one point, called its centre of gravity; the weight of the object can be considered to act at its centre of gravity. When an object is suspended from a point, its centre of gravity will be vertically below the point of suspension. A plumb line (weight attached to a piece of thread) is a useful way of showing the line of the vertical direction.

The position of the centre of gravity of an object affects its stability and how easy it will be to topple.

Aim

To find the centre of gravity of irregularly and regularly shaped plane lamina.

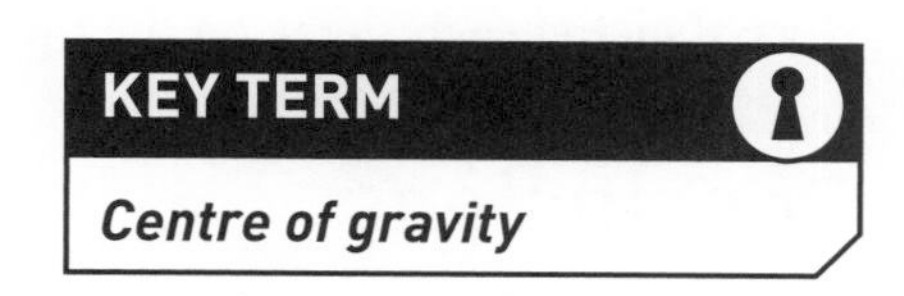

KEY TERM

Centre of gravity

Apparatus

- Retort stand and clamp
- Nail
- Regularly and irregularly shaped cardboard sheets (plane lamina)
- Small weight attached to a piece of thread
- Ruler
- Hole punch

SAFETY GUIDANCE

This practical presents minimal risk.

Method

In this practical you will suspend a plane lamina (thin sheet) from different points and determine the vertical line on which the centre of gravity lies.

Set up the apparatus as shown in Figure 1.

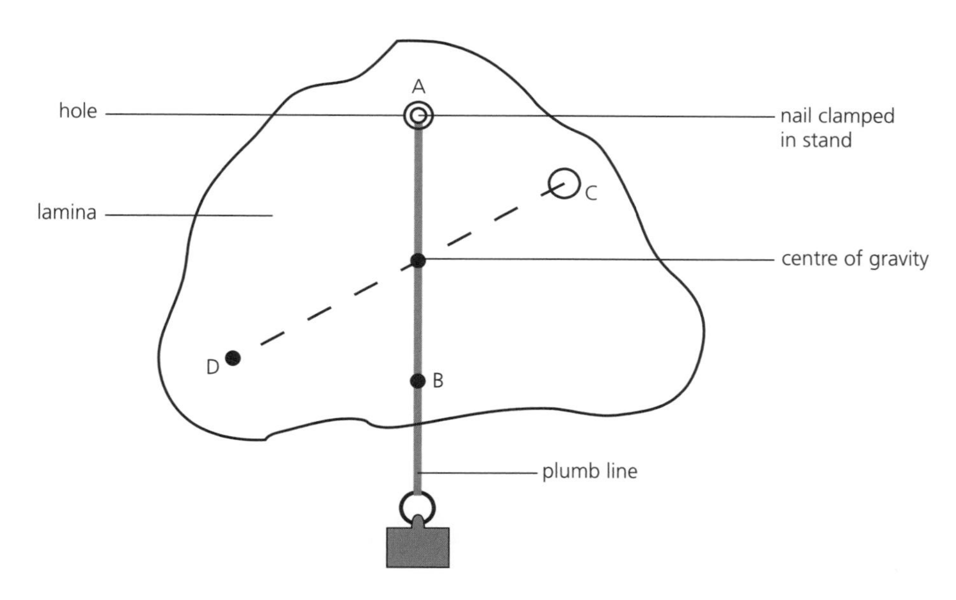

Figure 1

1 Use a hole punch to make a hole (A) in the irregularly shaped piece of cardboard and suspend it from a nail clamped in the retort stand.

2 With the shape at rest, suspend a plumb line from the nail and mark its position on the cardboard at three or four points. Remove the cardboard from the nail and draw the line AB by joining the points with a ruler.

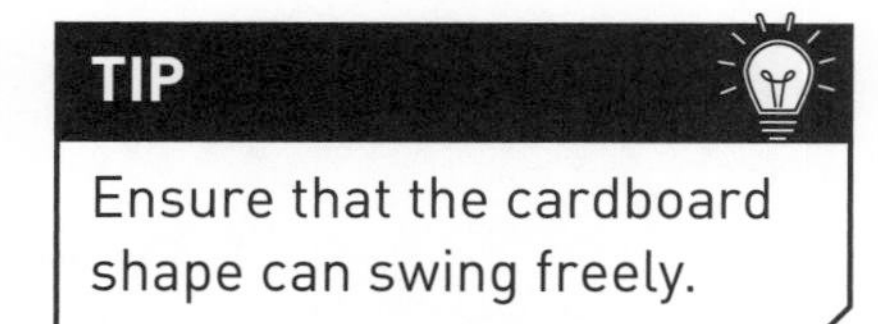

TIP

Ensure that the cardboard shape can swing freely.

3 Make a second hole (C) in the cardboard and repeat step **2** with the shape now suspended from C. Draw the line CD.

4 Make a third hole (E) in the cardboard and check that when the shape is suspended from E, the plumb line passes through the intersection of the lines AB and CD.

5 Repeat steps **1** to **4** with a regularly shaped piece of cardboard.

Observations

Mark and label the position of the centre of gravity on each of the cardboard shapes.

Try balancing each shape on the tip of your first finger.

Conclusions

1 State where you had to place each shape to make it balance on the tip of your finger.

....

2 The centre of gravity of a regularly shaped object that has the same density throughout is at its centre.

Locate and mark the centre of gravity in the following laminae.

a rectangle

b circle

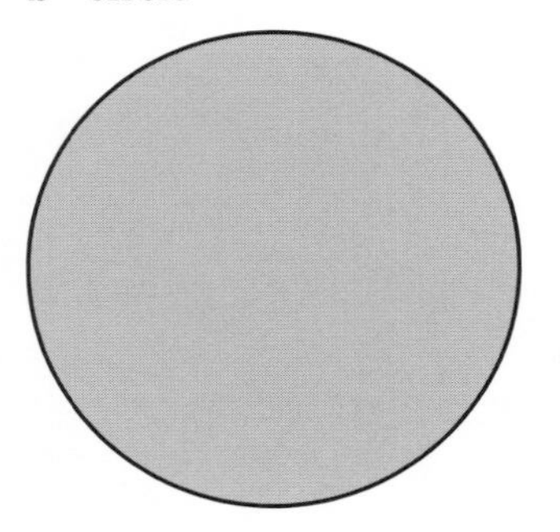

c equilateral triangle

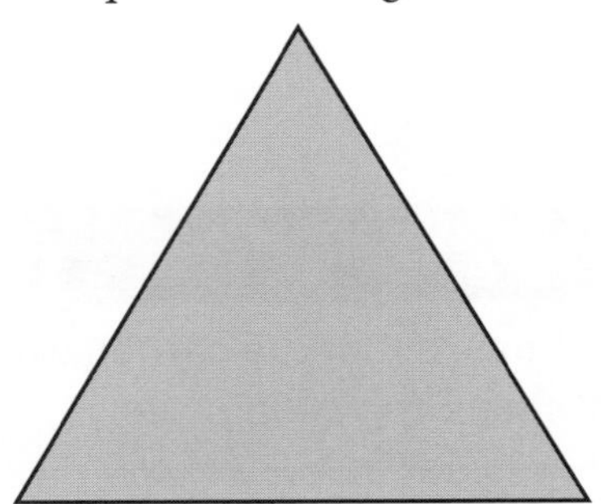

Evaluation

1 You may have noticed that the three lines you drew on your laminae did not intersect exactly at a point. State some possible sources of error in the experiment.

....

....

....

....

....

....

....

2 Suggest further precautions you could take to locate the centre of gravity more precisely.

....

....

....

....

....

....

....

....

....

1.7 Pressure

Pressure is defined as force per unit area. The greater the force and the smaller the area over which the force acts, the greater the pressure exerted. This property is widely used in hydraulic machines where a small force transfers pressure through a liquid to lift a heavy load. In this experiment you will discover how the pressure exerted on a solid object changes when the force applied is spread over areas of larger and smaller size.

Aim

To calculate the pressure you exert on the ground.

KEY TERMS 

Pressure
Accuracy
Gravitational field strength
Significant figures

Apparatus

- Bathroom scales
- Graph paper

KEY EQUATION

Pressure is given by:

$p = F/A$

where F is the force acting on area A.

SAFETY GUIDANCE 

This practical presents minimal risk.

Method

1 Stand on some bathroom scales and record your mass in kg.

2 Stand on a piece of graph paper and draw around the outline of one foot.

3 Count the number of squares on the graph paper inside the outline of your foot. Use this to estimate the contact area this foot has with the ground.

TIP 

If you do not wish to measure your own body mass, use a value such as 55 kg.

Observations

TIPS 

Remember that the area of a 1 cm × 1 cm square is 1 cm².

The outline of your foot may not always cover a square completely, so use the following rule: if half or more of a square is covered by your foot, include the square in your count; if less than half of a square is covered, do not count the square.

Mass of person = ... kg

Assuming the gravitational field strength, g, is equal to 10 N/kg, calculate the force you exert on the ground:

F = ... N

Number of squares covered by one foot = ...

Area of one foot = ... cm²

Convert the area from cm² to m²:

Area of one foot in m² = ...

Calculate the pressure on the ground, p_1, when you stand on one foot:

p_1 = ... N/m²

p_1 = ... Pa

State the contact area your two feet have with the ground: .... m²

Calculate the pressure on the ground, p_2, when you stand on two feet:

p_2 = ... N/m^2

p_2 = ... Pa

Conclusions

Summarise your results for standing on one foot and standing on two feet.

....

....

....

....

Evaluation

1 Outline how the precision of the experiment to measure the pressure exerted when you stand on both feet could be improved.

....

....

....

....

....

....

2 Justify the number of significant figures given in your answers.

....

....

....

....

GOING FURTHER

A simple barometer that measures atmospheric pressure is shown in Figure 1. The pressure at X due to the column of mercury XY equals the atmospheric pressure on the surface of the mercury in the bowl. There is a vacuum at Y. At a depth Δh below the surface of a liquid, the change in pressure is given by

$\Delta p = \rho g \Delta h$

where ρ is the density of the liquid and g is the strength of the Earth's gravitational field.

The height XY measures the atmospheric pressure in mm of mercury (mm Hg).

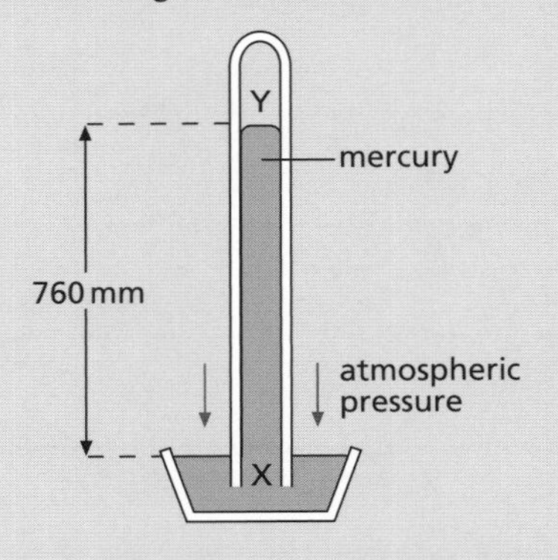

Figure 1 Mercury barometer

Explain whether the reading on the barometer would change if the tube was narrower or tilted.

1 Narrower

....

....

....

....

2 Tilted

....

....

....

....

2 Thermal physics

2.1 Specific heat capacity

The specific heat capacity of a material is defined as the energy required per unit mass per unit temperature increase. Different materials have different values of specific heat capacity. Materials with a high value require transfer of more energy to raise their temperature by the same amount than those with a low value if the mass of each material is the same. For example, different cooking pots will heat up (or cool down) more quickly, due to their different specific heat capacities.

KEY TERMS

Energy transfer
Temperature
Significant figures

KEY EQUATIONS

The specific heat capacity of a material, c, is given by:

$$c = \frac{\Delta E}{m\Delta\theta}$$

where ΔE is the energy required to raise the temperature of a mass, m, of the material by an amount $\Delta\theta$.

Power is given by energy transferred per unit time:

$$P = \frac{\Delta E}{t}$$

where ΔE is the energy transferred in time t by a device of power P.

Aim

To measure the specific heat capacity of a liquid and a solid.

Apparatus

- Eye protection
- Thermometer
- Electric immersion heater (12 V, 50 W)
- 12 V power supply and connecting leads
- Metal calorimeter or large metal pan
- Solid metal (aluminium) cylinder (with one central hole and one other hole)
- Heat-proof mat
- Oil
- Water
- Balance
- Timer

SAFETY GUIDANCE

- Eye protection must be worn.
- **Warning!** When using hot liquids and solids, ensure that you set them in a safe position where they will not be knocked over accidentally. Handle them with caution to avoid burns.
- Do not completely immerse the heater in water; the connecting cables must be above the water at all times.
- Do not switch on the heater when it is not in the water.

Method

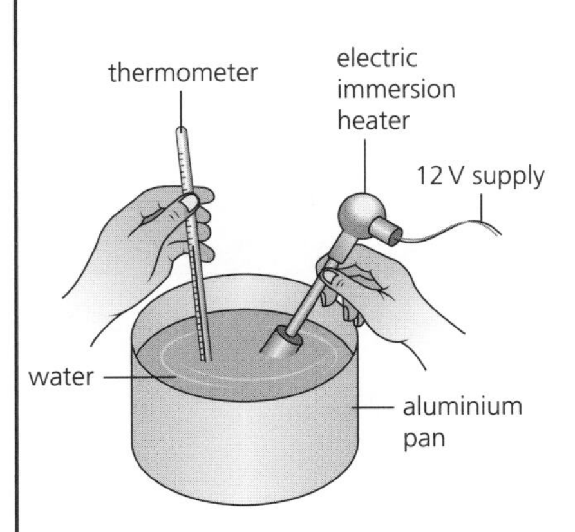

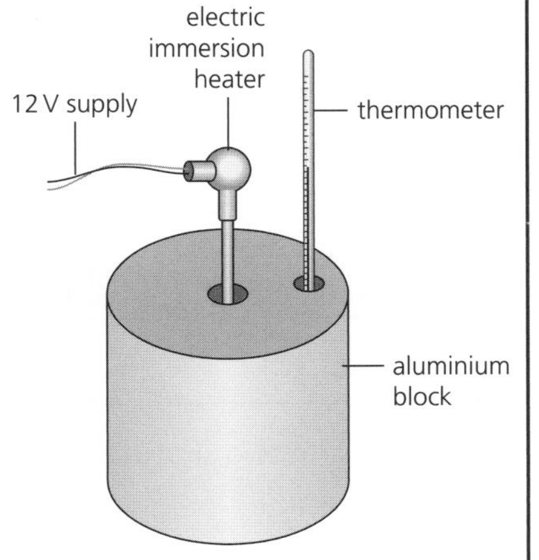

Figure 1

Liquid

1 Place an empty calorimeter on the balance and add 1.0 kg water.
2 Set the calorimeter on the heat-proof mat.
3 Record the temperature of the water, θ_1, in Table 1.
4 Insert the immersion heater into the water.
5 Switch on the heater and start timing.
6 Stir the water and after 5 minutes switch off the heater; record the heating time.
7 Continue stirring the water and note the highest temperature reached, θ_2.
8 Record the power of the immersion heater.

TIP 

After switching off the heater, make sure the temperature has stopped rising before you record its value.

TIP 

Check the power with your teacher if it is not marked on the immersion heater.

Solid

1 Find the mass of the metal cylinder, in kg. Record this in the Observations section.
2 Set the cylinder on a heat-proof mat. Place the immersion heater in the central hole of the cylinder and the thermometer in the other hole.
3 Record the temperature, θ_1, of the cylinder in Table 2.
4 Switch on the immersion heater and start timing.
5 After 5 minutes turn off the heater; record the heating time. Do not touch the metal block until it has cooled down.
6 When the temperature stops rising, record the highest temperature, θ_2, reached by the thermometer.
7 Record the power of the immersion heater.

TIP 

A little oil poured into the holes will improve thermal contact of the heater and thermometer with the cylinder.

Observations

Liquid

1 Power of heater, P = ... W

Mass of water, m = ... kg

Table 1

Initial temperature θ_1/°C	Highest temperature θ_2/°C	Change in temperature $\Delta\theta$/°C	Heating time t/s	Energy supplied $\Delta E = Pt$/J

2 Calculate the specific heat capacity of water (to two significant figures) using the equation:

$$c = \frac{\Delta E}{m\Delta\theta}$$

Specific heat capacity of water = ... J/(kg °C)

Solid

1 Power of heater, P = ... W

Mass of metal cylinder, M = ... kg

Table 2

Initial temperature θ_1/°C	Highest temperature θ_2/°C	Change in temperature $\Delta\theta$/°C	Heating time t/s	Energy supplied $\Delta E = Pt$/J

2 Calculate the specific heat capacity of the metal cylinder (to two significant figures).

Specific heat capacity of metal cylinder = ... J/(kg °C)

Conclusions

Summarise your results for the specific heat capacities of the liquid and solid.

....

....

....

....

Evaluation

1 Describe and give reasons for the precautions you took to ensure that the temperature rise of the water or metal block were measured accurately.

....

....

....

....

....

....

....

2 Compare your results with the expected values for the specific heat capacity of water and aluminium. These can be found from an internet or library search. Do your results agree with the published data? If not, suggest reasons why your experiments produced different values.

....

....

....

....

....

....

....

....

....

3 If you were to repeat the experiments, describe the improvements you would make to reduce sources of error.

....

....

....

....

....

GOING FURTHER

1 Aluminium has a higher specific heat capacity than copper. If two identical cylinders of copper and aluminium are heated to the same temperature and then allowed to cool, which would you expect to cool fastest, and why?

....

....

....

....

2 Suggest, in terms of its specific heat capacity, why water is used in the radiators of central heating systems.

....

....

....

2.2 Cooling curves

When an object cools, energy is transferred to the environment. When the object has cooled to the same temperature as its environment, equilibrium is reached and there is no net transfer of energy between the object and its environment. The rate at which energy is transferred from the object depends on a number of factors including its temperature, surface area, volume and the environmental conditions. You will investigate the dependence of the rate of cooling on some of these factors in this experiment.

Aim

To investigate factors that affect the rate of cooling from the surface of a liquid.

Apparatus

- Eye protection
- Two beakers of different diameters (250 cm^3 and 500 cm^3)
- Insulating material (and rubber bands)
- Timer
- Electronic balance
- Hot water (80°C to 90°C)
- Measuring cylinder (100 cm^3 or 250 cm^3)
- Thermometer (–10°C to 110°C)

Method

1 Wrap the insulating material around the sides of the beakers.

2 Pour 200 cm^3 of hot water into a measuring cylinder and then transfer the water quickly to the 250 cm^3 beaker. Start the timer immediately.

3 Record the temperature of the water in the beaker every 30 seconds for 10 minutes.

4 Repeat steps **2** and **3** with 200 cm^3 of hot water in the 500 cm^3 beaker.

KEY TERM

Equilibrium

SAFETY GUIDANCE

- Eye protection must be worn.
- **Warning!** Set hot liquids and solids in a safe position where they will not be knocked over accidentally. Handle them with caution to avoid burns.

TIP

Hold the insulation in place with rubber bands.

TIPS

Warm the measuring cylinder and beaker with hot water before starting the experiment.

Keep the beaker away from draughts during cooling.

Observations

Complete Table 1.

Table 1

Time/s	θ_1/°C (small beaker)	θ_1/°C (large beaker)	Time/s	θ_1/°C (small beaker)	θ_1/°C (large beaker)

Plot graphs of temperature on the y-axis against time on the x-axis for both sets of cooling results.

TIPS 

Choose an easy to read/plot scale for your graphs, mark points clearly, label axes correctly (including units) and use as much of the graph paper as possible.

Draw the best smooth curve through your data points.

Conclusions

Use your graphs to answer the following questions.

1 Determine the beaker that allows the water to cool most quickly. Justify your answer by reference to your results.

....

....

....

2 State whether the water cooled more quickly at higher or lower temperatures. Give reasons for your answer.

....

....

....

....

Evaluation

1 Suggest a reason why the measuring cylinder and beakers were both warmed with hot water before starting the experiment.

....

....

....

2 Explain why the beakers were kept away from draughts in the experiment.

....

....

....

3 What factors do your results suggest may influence the rate of cooling of the water in the beakers?

....

....

....

4 State the variables you tried to control in the experiment.

....

....

....

5 Can you suggest an improvement to the experimental method?

....

....

....

....

GOING FURTHER

1 Calculate the ratio of the surface area (A) to the volume (V) for a sphere of radius r.

2 It is suggested that the rate of cooling of an object increases as the ratio A/V increases.
If this is true, would you expect a sphere of large radius to cool faster or slower than a sphere of the same material but of smaller radius? Give a reason for your answer.
Assume the spheres start cooling at the same temperature and exist in the same environment.

....

....

....

....

2.3 Conduction and radiation

Thermal energy can be transferred in different ways. For conduction, matter must be present (a solid, liquid or gas). Some materials such as metals are good thermal conductors and the transfer of energy occurs quickly. Insulators such as wood are poor thermal conductors and transfer of energy occurs only very slowly.

For radiation, there is no need for matter to be present. All objects emit infrared radiation. Some colours and surfaces are better emitters of infrared radiation than others so they cool down more quickly.

Aim

To demonstrate the properties of good thermal conductors and good emitters of infrared radiation.

KEY TERMS 

Conductor
Insulator
Infrared radiation
Emitter

Apparatus

- Eye protection
- Selection of rods of different materials with the same dimensions
- Candle and matches
- Bunsen burner
- Tripod
- Timer
- Two identical thermometers (one with bulb painted dull black)
- Beaker of hot water
- Two retort stands with clamps
- Card

SAFETY GUIDANCE 

- Eye protection must be worn.
- **Warning!** Set hot liquids and solids in a safe position where they will not be knocked over accidentally. Handle them with caution to avoid burns.
- Take care when handling the hot candle wax.

Method

Conduction

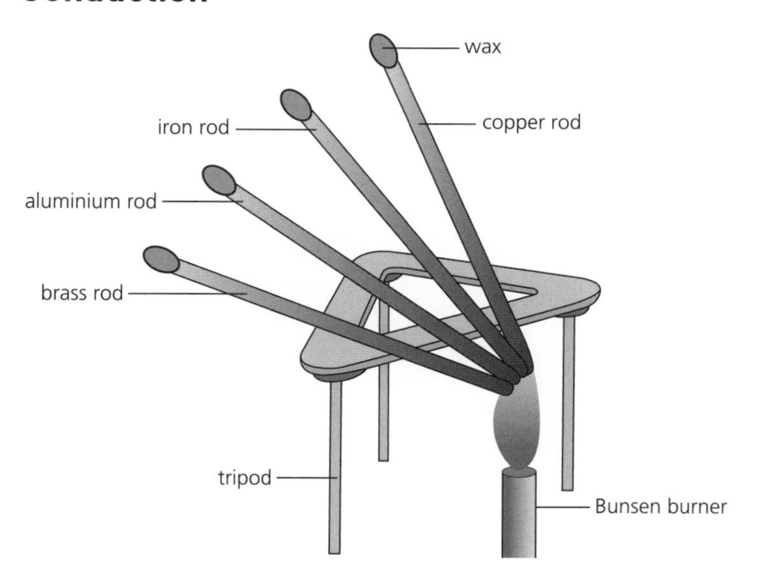

Figure 1

1 Melt a small blob of candle wax onto one end of each rod.

2 Rest the rods on a tripod and arrange them in a fan shape such that the unwaxed ends of each rod are close together and the waxed ends are separated. Place a card under the waxed ends to catch the drips of wax!

TIP 

Apply the same small amount of wax to each rod.

3 Start the timer and heat the unwaxed end of the rods evenly and gently with a Bunsen burner.

4 Record the time at which the wax starts to drip from each of the rods.

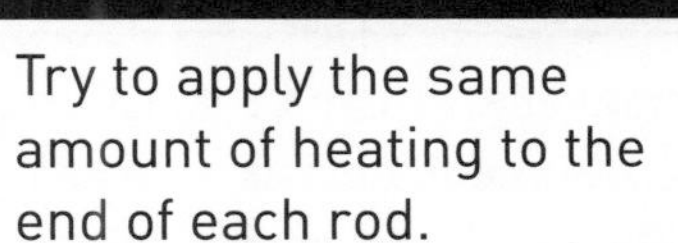

Radiation

1 Select one thermometer with an unpainted bulb and one with a bulb painted black.

2 Record the room temperature.

3 Place the thermometers in a beaker of hot water and wait until they reach the same steady temperature; record this temperature.

4 Remove the thermometers from the water at the same moment and start the timer.

5 Clamp each thermometer in a retort stand.

6 Record the temperature of each thermometer every 30 seconds for 10 minutes.

7 Plot a graph to show how the temperature reading on each thermometer varied with time.

TIP

Work in pairs with each student recording the temperature of one of the thermometers.

TIP

Place the thermometers at least 20 cm away from each other and from the hot water, and also away from any draughts.

Observations

Conduction

1 How is thermal energy transferred along the metal rods?

....

2 Complete Table 1, including any units in the column headings.

Table 1

Material	Time taken by wax to melt/

Radiation

1 Measure the room temperature and note it below. Complete Table 2.

Room temperature = ...°C

Table 2

Time/s	Temperature of thermometer with black bulb/°C	Temperature of thermometer with shiny bulb/°C

Time/s	Temperature of thermometer with black bulb/°C	Temperature of thermometer with shiny bulb/°C

2 Plot a graph of temperature on the y-axis and time on the x-axis for each thermometer.

Conclusions

1 a How can you tell which material is the best thermal conductor?

....

....

b List the materials in order of conductivity from best to worst.

....

....

....

....

2 a Which thermometer cools down the fastest?

....

b Complete the following sentence.

A dull black surface emits infrared radiation .... a shiny surface.

Evaluation

Conduction

1 State the variables which were controlled.

....

....

2 Suggest sources of error in the conduction experiment.

....

....

....

Radiation

1 Explain why the two thermometers should have the same temperature when removed from the hot water.

....

....

....

2 Why should the thermometers be shielded from a draught?

....

....

....

3 State the variables which were controlled.

....

....

4 Suggest sources of error in the radiation experiment.

....

....

....

....

....

GOING FURTHER

Discuss whether radiators in a central heating system should be painted black or white.

....

....

....

....

....

3 Waves

3.1 Ripple tank experiments (Teacher demonstration)

A ripple tank can be used to study the reflection, refraction and diffraction of waves.

It consists of a transparent tray containing water to a depth of about 5 mm. A light source casts shadows of the water waves onto a white screen.

Looking through a stroboscope – a disc with a number of equally spaced slits in it – makes it easier to study the wave motion. When the disc rotates at a speed such that each time a view of the screen is obtained the waves have advanced one wavelength, the waves appear to be 'frozen' or stationary.

Aim

To study the general properties of waves using a ripple tank.

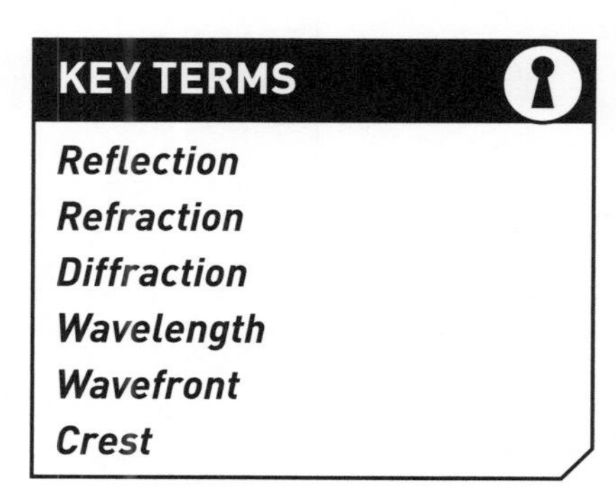

Apparatus

- Ripple tank and accessories (including vibration generator (motor) and power supply)
- Large sheet of white paper
- Hand stroboscope
- Metre ruler
- Straight metal strips
- Rectangular glass plate

SAFETY GUIDANCE 

- People with epilepsy can be affected adversely by stroboscopic lighting. Before using a stroboscope, check that no one in the class is likely to be affected.
- Mop up any spilled water immediately to prevent slips and falls.

Method

Although your teacher may set up the demonstration, you should still read through the method to ensure you understand what is happening. You will be asked to make sketches of the wavefronts on the white paper viewed through a stroboscope and answer questions about them.

Ripples are produced by dipping an object into the water repeatedly. If a straight bar is attached to an electric motor, as shown in Figure 1, continuous ripples are generated. Circular ripples are generated if the bar is replaced by a small ball.

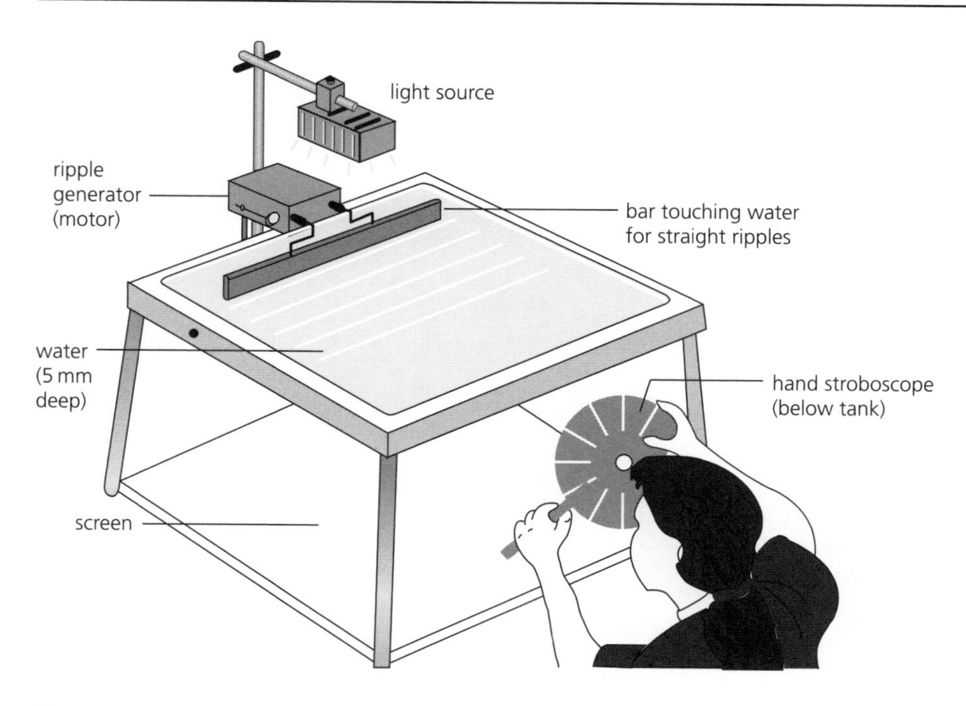

Figure 1

Reflection

1. Generate some continuous straight waves in the ripple tank.
2. Place a straight metal strip in the tank at an angle of about 60° to the wavefronts and observe the reflection of the waves.
3. Adjust the speed of rotation of the stroboscope until the water waves appear to be stationary (frozen).
4. Make a sketch of the shadows obtained on the screen; the wavefronts are represented by the straight lines and can be regarded as the crests of waves.
5. On your sketch mark the change of direction of the wavefronts when they are reflected from the metal strip.

Refraction

1. Place a rectangular glass plate in the ripple tank; align it so that a straight edge is about 45° to the wavefronts. The plate should be of a thickness such that the depth of water above it is about 1 mm; the water depth is around 5 mm elsewhere in the tank.
2. Generate some continuous straight waves.
3. Adjust the speed of rotation of the stroboscope until the water waves appear to be stationary (frozen).
4. Make a sketch of the shadows obtained on the screen; the distance between the straight lines corresponds to the wavelength (the distance between wave crests) of the wave.
5. Estimate the wavelength of the waves in the deep and shallow regions.

TIP 

Mark the distance occupied by multiple (N) wave crests and calculate an average wavelength by dividing the distance by (N – 1).

Diffraction

1 Place an obstacle in the ripple tank. The obstacle should have a gap in it of about the same width as the wavelength of the water waves (Figure 2).

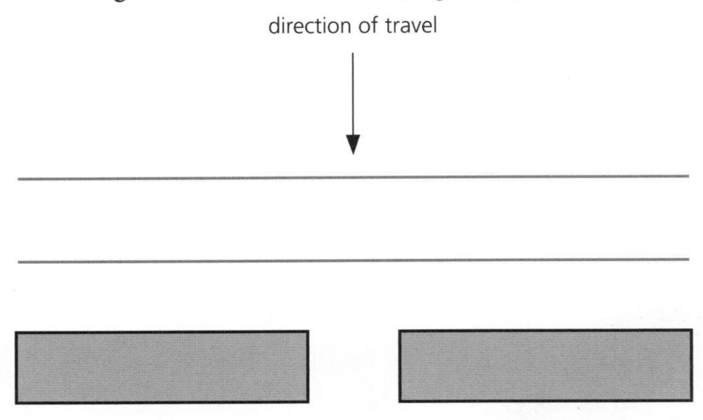

Figure 2

2 Make a sketch of the shadows obtained on the screen when continuous straight waves are incident on the gap.

3 Replace the obstacle with one having a much wider gap. Make a sketch of the image obtained on the screen when continuous straight waves are incident on the gap.

Observations

Reflection

In the space below, sketch the wavefronts incident on and reflected from the metal strip.

Refraction

In the space below, sketch the wavefronts passing over different depths of water.

Approximate value of:

Wavelength of waves in deep water = ...

Wavelength of waves in shallow water = ...

Diffraction

In the space below, sketch the wavefronts passing through the narrow gap.

Sketch the wavefronts passing through the wide gap.

Conclusions

1 State what happens to the water waves when they reach the metal strip.

....

2 State whether the wavelength is shorter or longer in the shallow water.

....

3 State whether the water waves travel faster or slower in shallow water.

....

4 Describe what happens to the direction of travel of the waves as they enter shallow water.

....

....

5 Describe what happens to the waves when they pass through a narrow gap.

....

....

6 Describe any differences you notice in the image of the waves passing through a wide gap compared to the image produced when the waves pass through a narrow gap.

....

....

Evaluation

1 Explain the advantage of using a stroboscope to estimate the wavelength of the waves.

....

....

2 Suggest one way in which the experiment could be improved.

....

....

....

GOING FURTHER

When an earthquake occurs under the sea floor, the energy released can be transmitted to the surrounding water causing a tsunami. In water of constant depth, this spreads out as a circular wave. A tsunami wave can travel for very long distances across the ocean. Scientists can predict the time of arrival of a tsunami in different parts of the world if the distance from the epicentre of the earthquake and the speed of travel of the wave are known. The speed of travel varies with the depth of water across which the wave travels.

1 Estimate the average speed of the tsunami that took about two hours to travel from Indonesia to Sri Lanka, a distance of 1460 km, after the 2004 Sumatra–Andaman earthquake.

2 The tsunami took about the same time (two hours) to travel from Indonesia to Thailand, a much shorter distance than to Sri Lanka. What does this suggest about the depth of water between Indonesia and Thailand compared to the depth of water between Indonesia and Sri Lanka?

....

....

....

3.2 Reflection in a plane mirror

When a light ray reflects from a surface, the angle of incidence is the angle between the incident ray and the normal to the surface. The angle of reflection is the angle between the reflected ray and the normal to the reflecting surface. The law of reflection states that the angle of incidence equals the angle of reflection.

The image formed by an object in a plane mirror is always virtual, meaning that it cannot be formed on a screen. You will learn more about the location and characteristics of the image in this practical work.

Aim

To verify the law of reflection and identify the properties of an image in a plane mirror.

KEY TERMS

Angle of incidence
Angle of reflection
Image
Normal
Ray

Apparatus

- Lamp
- Single slit
- Plane mirror
- Optics pins
- Protractor
- Blu-tack or adhesive/mounting putty
- 30 cm ruler
- Cork mat

Method

Investigating the law of reflection

1 On Figure 2, use a protractor to draw in lines from O at 15°, 30°, 45°, 60° and 75° to the normal ON (as indicated in Figure 1).

2 Place a plane mirror vertically on the line AOB; blu-tack may be helpful to hold it in position.

3 Align the lamp and single slit so that a beam of light is incident on the mirror along the 15° line; use a ruler to mark the position of the reflected ray.

4 Repeat step **3** for the other angles of incidence.

5 Remove the mirror and measure the angle of reflection for each ray.

6 Record your results in Table 1.

KEY EQUATION 

Law of reflection:

$i = r$

where i is the angle between the incident ray and the normal to the reflecting surface (the angle of incidence) and r is the angle between the reflected ray and the normal to the reflecting surface (the angle of reflection).

SAFETY GUIDANCE 

Take care when using the lamp, it can get very hot.

TIP 

The reflecting surface of the mirror must lie exactly on line AOB.

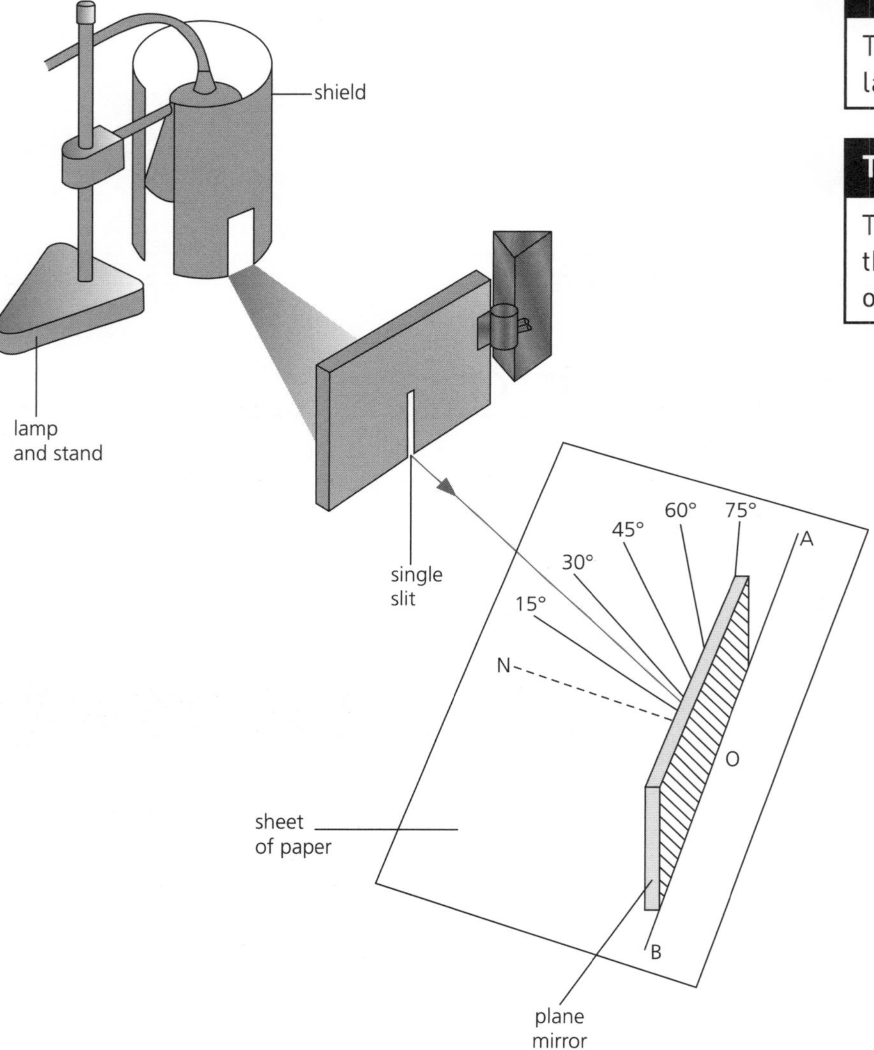

Figure 1 Experimental apparatus used to investigate the law of reflection, $i = r$

Locating the position of the image in a mirror

1 Replace the mirror on line AOB.

2 Mark a point about 3 cm from the mirror on the 30° incidence line; set a pin vertically on this point, P_O.

3 Locate the image of the pin in the mirror by looking along the 30° reflection line; mark the apparent position, P_I, of the image behind the mirror with a second pin. When you move your head up and down, the image of the first pin and the top of the second pin behind the mirror should coincide; when you move your head parallel to the mirror they will remain coincident if you have located the position of the image P_I correctly.

4 Measure and record the perpendicular distances of P_O and P_I from the line AOB.

TIP

A cork mat placed under the paper will help to hold the pin in place.

TIP

Have your eyes at table level when you are locating P_I.

Observations

Investigating the law of reflection

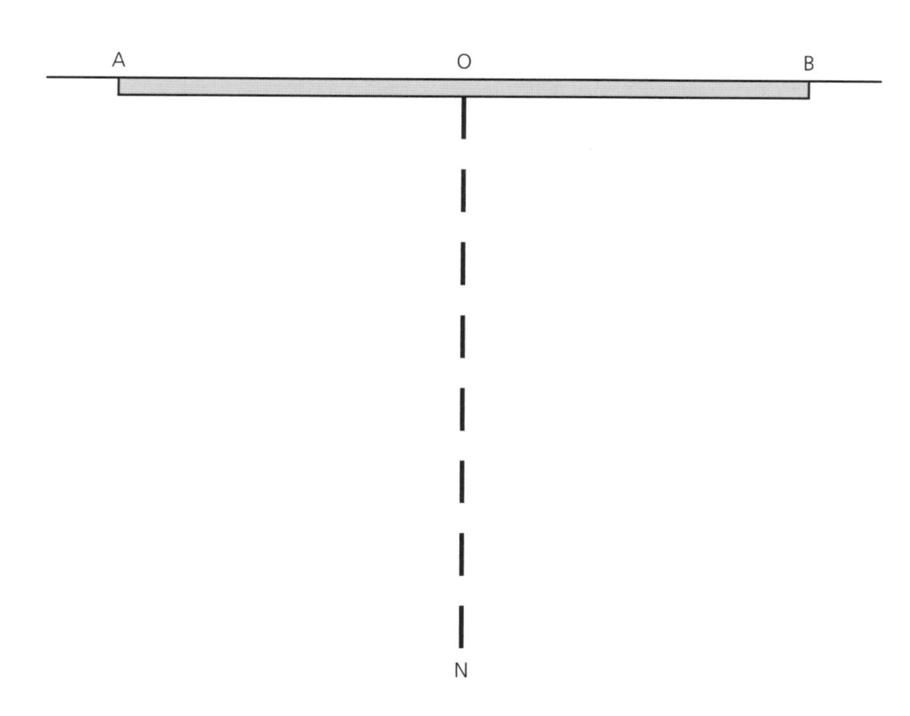

Figure 2

Record your measurements in Table 1.

Table 1

Angle of incidence, i/°	Angle of reflection, r/°

Locating the position of the image in a mirror

Along a line joining P_O and P_I:

The perpendicular distance of P_O from mirror = ...

The perpendicular distance of P_I from mirror = ...

Conclusions

1 State the relationship between the angle of incidence and the angle of reflection. Use data from your experiment to justify your answer.

....

....

2 Describe where the image in a plane mirror is formed. Use data from your experiment to support your answer.

....

....

....

3 State the characteristics of the image formed in a plane mirror.

....

....

....

Evaluation

1 Explain why it is important to align the reflecting surface of the mirror exactly along line AOB.

....

....

....

2 State the precision of measurements taken with a protractor.

....

3 Suggest some sources of error when measuring the angles of incidence and reflection.

....

....

....

....

....

4 Suggest some sources of error when locating the position of the image in the mirror.

....

....

....

GOING FURTHER

1 Suggest what is meant by lateral inversion.

....

....

....

....

....

2 How could you test whether the image in a plane mirror is laterally inverted?

....

....

....

....

....

....

....

3.3 Refraction of light

When light passes from air into a different medium (such as glass or water) at an angle of incidence greater than zero, it is bent or refracted at the boundary between the two media. If the speed of light is less in the second medium than in air, light is refracted towards the normal to the surface. On emerging from the second medium and returning to air, refracted light is bent away from the normal.

The refractive index of a transparent material gives a measure of the amount of refraction that occurs, in terms of the sines of the angle of incidence and angle of refraction at the boundary.

Aim

To observe the path of light rays passing through a glass block and measure the angle of refraction for different angles of incidence.

To use these measurements to determine the refractive index of glass.

KEY TERMS 

Angle of incidence
Angle of refraction
Normal
Refractive index

KEY EQUATION 

The refractive index, n, of a medium is given by:

$$n = \frac{\sin i}{\sin r}$$

where i is the angle between the incident ray in air and the normal, and r is the angle between the refracted ray and the normal in the medium.

Apparatus

- Lamp
- Single slit
- Glass block with lower surface painted white
- Protractor
- 30 cm ruler
- Calculator

Method

You will investigate the refraction of light in a rectangular glass block. Before you take any measurements, read the instructions below. Complete Table 1 to identify which variable you will change, which variables you will keep fixed and which variable you will measure.

Table 1

Independent variable (variable you change)	Control variables (constant/fixed)	Dependent variable (variable you measure)

1 Place the glass block with the white painted surface on Figure 1 overleaf, with its long side along line AOB. Draw the outline of your block onto Figure 1 as it will not be exactly the same size as the rectangle in Figure 1.

2 Align the lamp and single slit so that a beam of light is incident on the block at O at an angle of incidence of about 30°. The ray of light will be refracted at the edge of the block, travel through the block and be refracted again when it leaves the block.

Draw two crosses along the path of the ray entering the block and two crosses on the ray emerging from the block. Mark the crosses as far apart as possible.

SAFETY GUIDANCE 

Take care when using the lamp, it can get very hot.

3 Take the block off the paper and use a ruler to join the two crosses for the incident ray, and the two crosses for the emergent ray; extend these lines back to the block. Connect the two lines to draw the path of the light refracted through the block.

Mark the normal to the block at point O.

Mark the normal to the block for the emergent ray.

Measure the exact angle of each ray to the normal with a protractor. Record your results in Table 2.

TIP 

Use a sharp pencil to draw ray paths.

4 Place the block back on the outline and repeat steps **2** and **3** with an angle of incidence of about 60° and 0°.

5 Move the lamp, slit and glass block to Figure 2. Place the glass block with its short side along the line AOC and draw the outline of your block on the paper.

6 Repeat steps **2** to **4**, recording your results in Table 3.

TIP 

Ensure the glass block remains exactly aligned on line AOB.

Observations

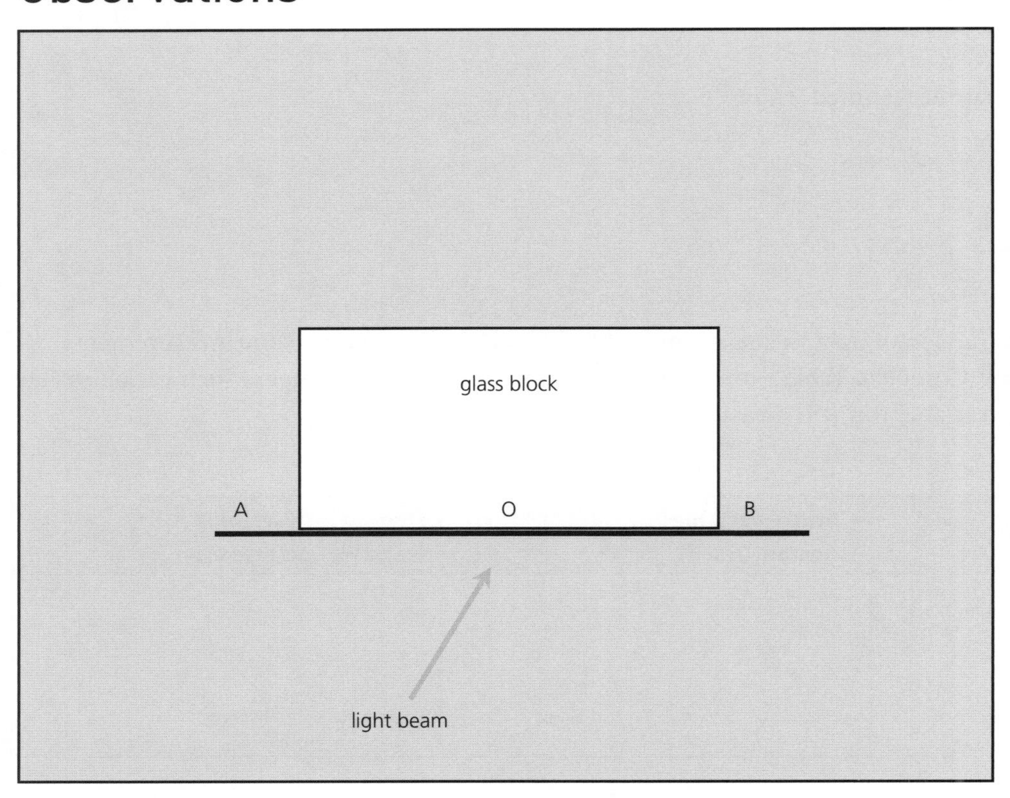

Figure 1

State the angle in degrees between the normal and the surface (AOB) of the glass block.

....

Table 2 Light beam incident on AB

Angle of incidence, i/°	Angle of refraction, r/°	Angle of emergent ray, θ/°

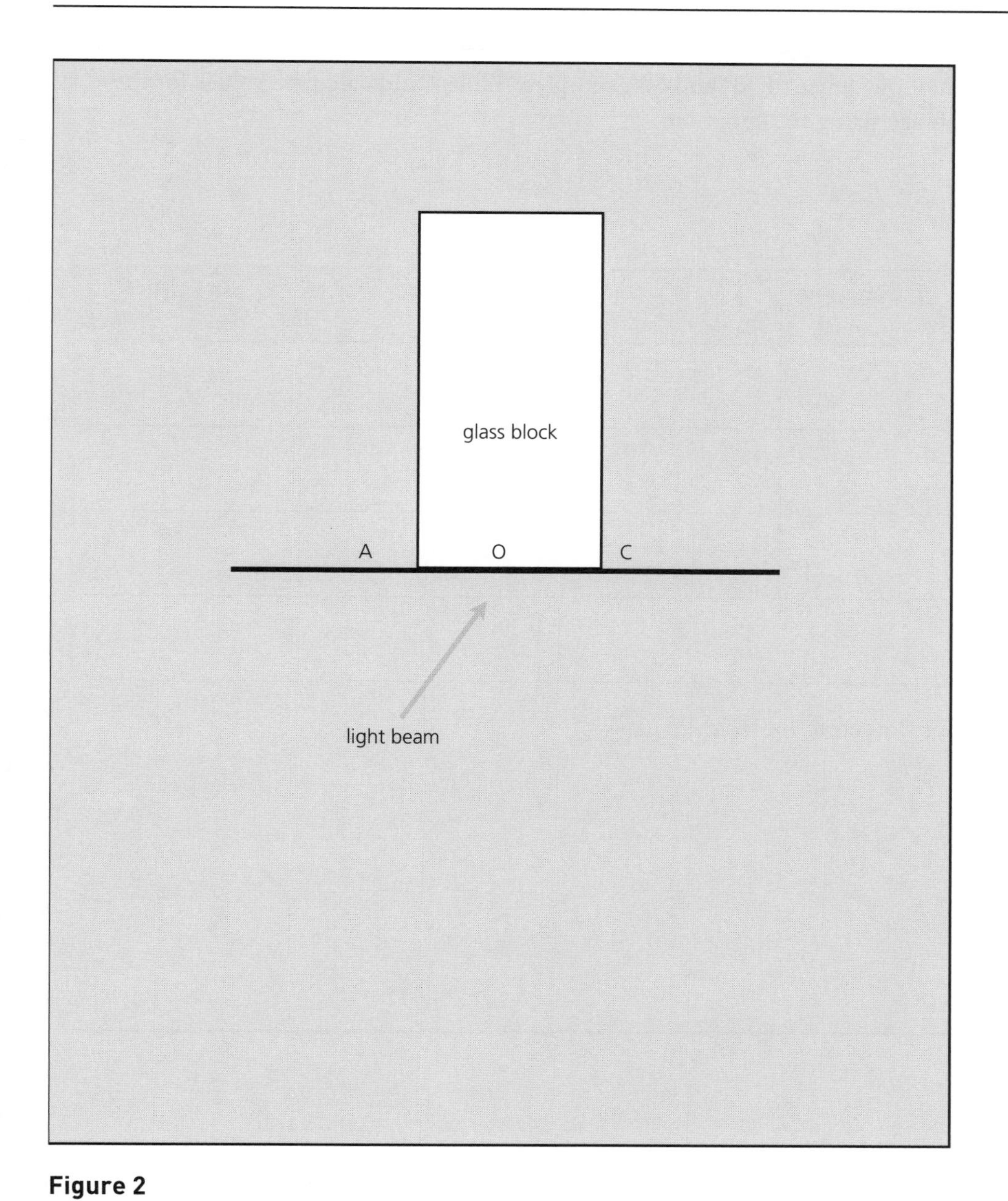

Figure 2

Table 3 Light beam incident on AC

Angle of incidence, i/°	Angle of refraction, r/°	Angle of emergent ray, θ/°

Using your results for angles of incidence of 30° and 60°, complete Table 4 and calculate values for the refractive index of the glass block using the equation:

$$n = \frac{\sin i}{\sin r}$$

Table 4

Angle of incidence, i/°	Angle of refraction, r/°	$\sin i$	$\sin r$	$n = \frac{\sin i}{\sin r}$

Calculate an average value for the refractive index of glass.

Conclusions

1 The average value for the refractive index of the glass block is ....

2 State the angles of incidence for which the light was refracted.

....

....

3 What do you notice about the directions of the light rays entering and emerging from the block? Use data from Table 2 or Table 3 to support your answer.

....

....

....

4 Complete the following sentences about the path of light rays travelling through the glass block using the words below.

angle of incidence *away from* *equals* *parallel* *straight through* *towards*

a A light ray is refracted .... the normal when it enters the glass block. A light ray is refracted .... the normal when it leaves the block.

b A ray incident normally on a glass block is not refracted but passes .... the block.

c When a light ray undergoes refraction in a glass block it emerges .... to the incident ray. The angle of emergence .... the ....

Evaluation

Explain why it is difficult to obtain an accurate measurement of the angles of the rays and suggest improvements you could make to remove these difficulties.

....

....

....

....

....

GOING FURTHER

The speed of light in air changes when the density of the air changes.

Explain why the Sun is visible at sunset for about two minutes after it has sunk below the horizon.

....

....

....

....

3.4 Images formed by a converging lens

Lenses are used in many optical instruments from cameras and telescopes to magnifying glasses and spectacles. It is necessary to know the focal length of a lens for any particular application. In this practical you will use the fact that rays from a distant object are nearly parallel to find the focal length of a converging lens. The image formed by a converging lens can be real or virtual depending on the position of the object relative to the principal focus of the lens. The size of the image (unchanged, enlarged or diminished) and whether it is upright or inverted also depend on the position of the object relative to the lens.

Aim

To determine the focal length of a converging lens and investigate the images formed by a converging lens.

Apparatus

- Small torch and stand
- Converging lens
- White card
- Paper
- Metre ruler
- Blu-tack or adhesive/mounting putty
- Thin (translucent) paper

KEY TERMS

Focal length
Converging lens
Principal focus (focal point)
Real image
Virtual image

SAFETY GUIDANCE

Warning! To avoid eye damage never look directly at the Sun or any intense light source.

Method

You will find the position and characteristics of the image formed by a converging lens when the position of the object relative to the lens is varied. First you need to find the focal length of the lens.

Focal length of a lens

1 Hold the lens up in the direction of a window and form a sharp (focused) image of a distant object on the wall or a piece of card.
2 Measure and record in the Observations section the distance between the lens and the card.
3 Repeat steps **1** and **2** and obtain an average value for the distance between the lens and the focused image. Since the light rays from a very distant object are nearly parallel, the distance of the image from the lens is equal to the focal length, f, of the lens.

Characteristics of the image for different positions of the object

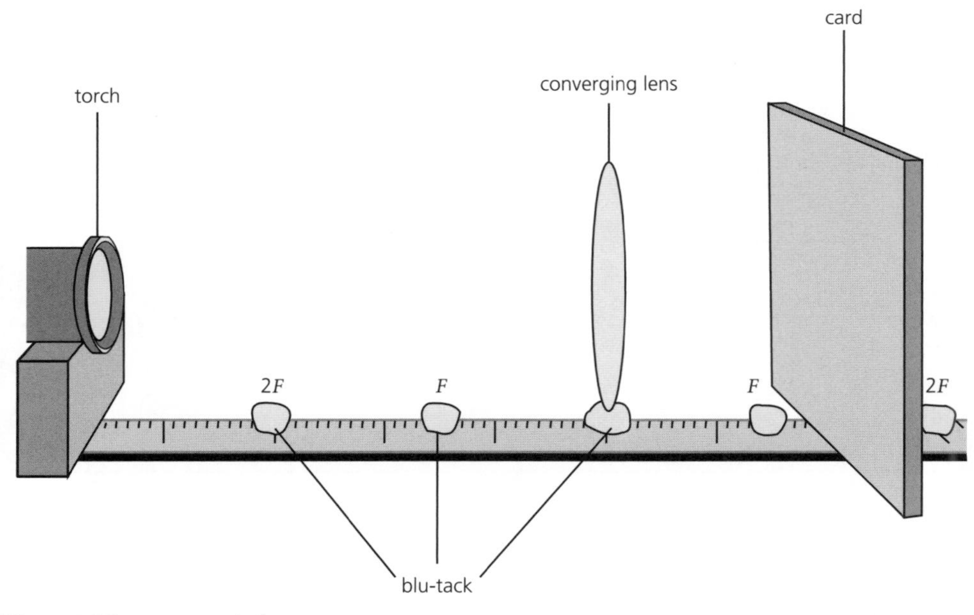

Figure 1 (not to scale)

1 Set up the apparatus as shown in Figure 1. Use some blu-tack to hold the lens vertically upright at the 50 cm mark of a metre ruler resting horizontally on the bench. Draw an arrow on a small piece of thin paper and attach this paper to the front of the torch to act as an object.

2 Use a little blu-tack to mark the position on the ruler of the principal focus, *F*, on each side of the lens. (The point *F* is a distance *f* from the centre of the lens; use the focal length you found in the first part of the investigation.) Also mark the 2*F* position each side of the lens (which is $2 \times f$ from the centre of the lens).

3 Set the torch beyond the 2*F* position facing the centre of the lens.

4 Measure and record in Table 1 the distance of the object from the lens.

5 Place the card on the opposite side of the lens to the object and move the card until you locate a sharp (focused) image of the object on the card.

TIP 

Set the torch so that the light from it strikes the centre of the lens.

TIPS 

Rotate the torch so that the arrow is upwards to allow you to identify the top of the image easily.

Mark the top and bottom of the image on the card to aid the height measurement.

6 Measure and record the distance of the image from the lens.

7 Measure and record the height of the image and whether it is inverted or not.

8 Move the torch nearer to the lens, first to 2*F*, then to between 2*F* and *F*, and repeat steps **4** to **7** for each position.

9 Move the torch to a position between *F* and the lens; it will not be possible to form an image on the card. Reduce the brightness of the torch (by placing a thin piece of paper in front of it) and look back through the lens towards the torch. Estimate and record the distance of the image from the lens and its height; record whether the image is upright or inverted.

10 Record the height of the object.

11 Using the equation

$$\text{magnification of an image} = \frac{\text{height of image}}{\text{height of object}}$$

find and record the magnification of the image.

Observations

Focal length of a lens

Image distance 1 = ...

Image distance 2 = ...

Average image distance = ...

Focal length, *f*, of lens = ...

Characteristics of the image for different positions of the object

Height of object = ... cm

Table 1

Torch position	Object distance/ cm	Image distance/ cm	Height of image/ cm	Magnification	Upright or inverted
beyond $2F$					
at $2F$					
between $2F$ and F					
between F and lens					

1 We found the height of the image by:

....

....

....

2 We determined whether the image was inverted or not by:

....

....

....

Conclusions

1 Summarise how the image distance, magnification and uprightness of the image varies as the object is moved towards the lens.

....

....

....

....

....

....

....

....

....

....

....

....

2 In which image positions is the image:

a real

....

b virtual?

....

3 Draw a ray diagram to show where the image is formed when the object lies between *F* and the lens.

Evaluation

Describe any problems that you had with the experiment. Suggest how these could be reduced or avoided to produce more accurate measurements.

....

....

....

....

GOING FURTHER

A simple telescope can be constructed with two converging lenses. The objective lens, used to form an image of a distant object, should have a long focal length; the eyepiece lens, used to magnify the image formed by the objective lens, should have a short focal length.

1 a i State where an image of a distant object is formed by the objective lens of the telescope.

....

ii Is the image real or virtual?

....

b i State where an image of a distant object is formed by the eyepiece lens of the telescope.

....

ii Is the image real or virtual?

....

2 You are supplied with two converging lenses of focal lengths 50 cm and 5 cm, a ruler, some blu-tack, a piece of greaseproof paper and a filament lamp. Describe how you could construct and test a model telescope using this apparatus.

....

....

....

....

....

....

....

....

....

3.5 Speed of sound in air

To measure the speed of sound you need to measure the time it takes for sound to travel a certain distance. The speed of sound is about ten times faster than the speed of a car travelling at 120 km/h, so for sound travelling a distance of a few metres, the time interval is only a few milliseconds. This is too small to be measured accurately by a person operating a stopwatch, so a timer is needed that is triggered to start and stop automatically. An automatic timer is more accurate because it eliminates reaction-time errors. It can record times to the nearest millisecond (0.001 s).

Aim

To measure the speed of sound in air.

Apparatus

- Electronic timer with millisecond precision
- Two microphones
- Small hammer
- Metal plate
- Metre ruler

KEY TERMS 

Speed
Microphone
Accuracy
Precision
Gradient
Significant figures

KEY EQUATION 

The speed of sound in air, v, is given by:

$$v = \frac{d}{t}$$

where d is the distance the sound travels in air in time, t.

Method

Set up the apparatus as shown in Figure 1, with the microphones about 1 metre apart. If other groups are also doing the experiment, set up your apparatus as far as possible from each other.

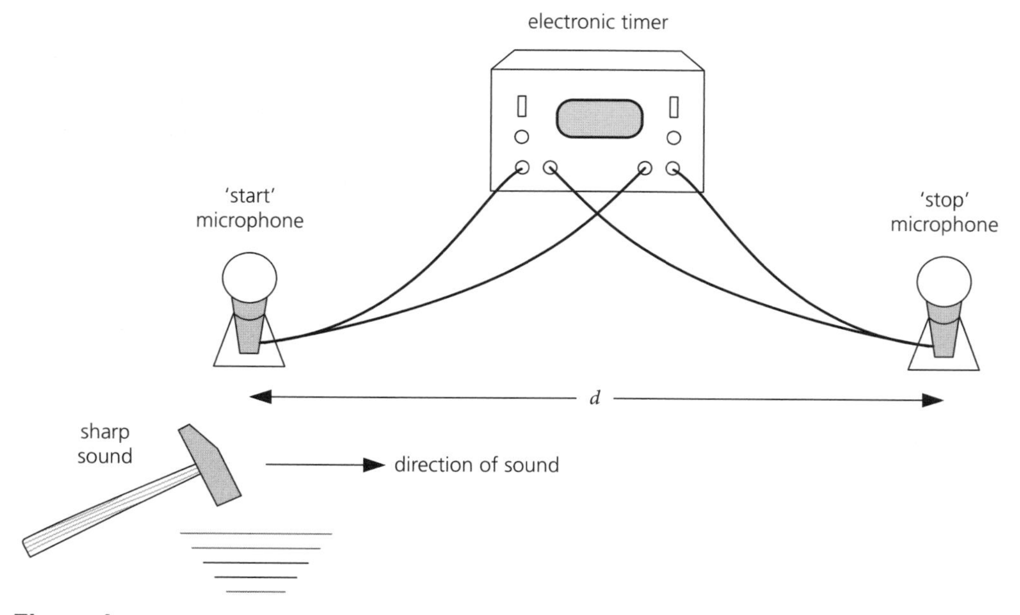

Figure 1

1. Measure and record the distance, d, between the centres of the 'start' and 'stop' microphones.
2. Switch on the timer so that it is ready to record.
3. Tap the hammer sharply on a metal plate placed close to the 'start' microphone; the timer should start.
4. The timer should stop when the sound reaches the 'stop' microphone.
5. Record the time, t_1, displayed on the timer.
6. Reset the timer and repeat steps **3** to **5**; record the time, t_2.
7. Move the microphones 10 cm closer to each other and repeat steps **1** to **6**.
8. Repeat step **7** three more times.

SAFETY GUIDANCE 

Take care when using the hammer.

TIP 

Check the units in which your timer displays time (remember that 1 ms = 0.001 s).

Observations

1 Complete Table 1.

Table 1

Distance, *d*/cm	First time measurement, t_1/ms	Second time measurement, t_2/ms	Average time, *t*/ms

2 Plot a graph of the average time, *t*, along the *x*-axis against the distance, *d*, along the *y*-axis. Draw a line of best fit.

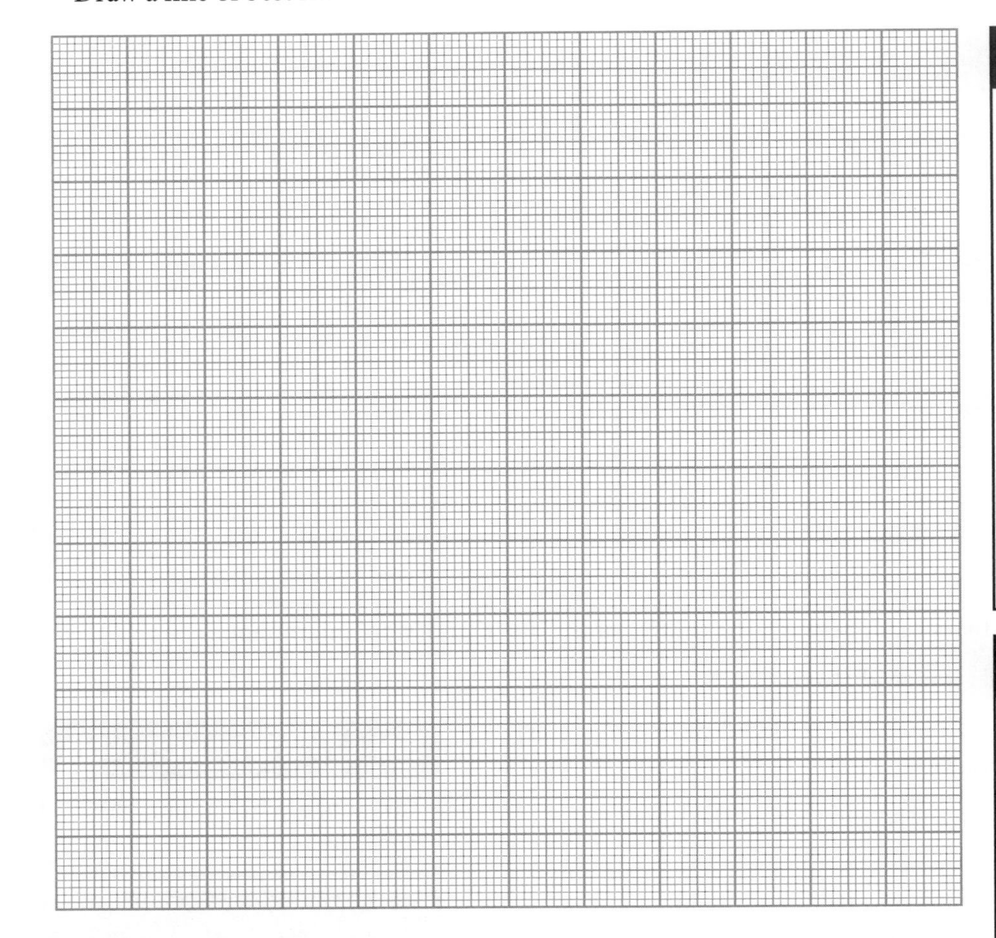

TIP 

Choose an easy to read/plot scale for your graph, mark points clearly and use as much of the graph paper as possible; axes should be labelled correctly with *d*/cm plotted vertically and *t*/ms horizontally. The best straight line should be drawn through the points with a sharp pencil.

TIP 

Use the triangle method (see page 13) to determine the gradient of your graph from as long a line as possible and show your working.

3 Calculate the gradient of the graph.

Summary of results:

a Gradient of graph = ... cm/ms

b Speed of sound in air = ... cm/ms

= .... m/s

> **TIP**
>
> To convert cm/ms to m/s multiply by the factor 1000/100.

Conclusions

State the value you obtained for the speed of sound in air to two significant figures. Compare your value with the expected value; suggest reasons for any difference.

....

....

....

....

....

Evaluation

1 **a** State the precision of measurements made with a ruler: ....

b Suggest why the accuracy of your measurement of distance *d* may be less than this.

....

....

....

....

c Estimate the accuracy of your measurements for *d*: ....

2 Explain why it was important to position your apparatus as far as possible from neighbouring groups.

....

....

....

....

3 Discuss any sources of error which could have affected your results and how the accuracy of the experiment could be improved.

....

....

....

....

....

....

GOING FURTHER

When a police car sounding its siren races past, you hear the note of the siren drop in pitch. When the police car moves away at high speed, state how the note of the siren appears to change in:

a frequency

....

b wavelength.

....

c Suggest why the changes occur.

....

....

....

3.6 Investigating pitch and loudness of sound waves (Teacher demonstration)

Sound waves can be turned into electrical signals with a microphone. These electrical signals can be displayed on a cathode ray oscilloscope (CRO). Unlike a voice or a musical instrument, a tuning fork emits a 'pure' note, i.e. a note of one frequency.

Aim

To use a cathode ray oscilloscope to display waveforms for sound waves of different pitch and loudness.

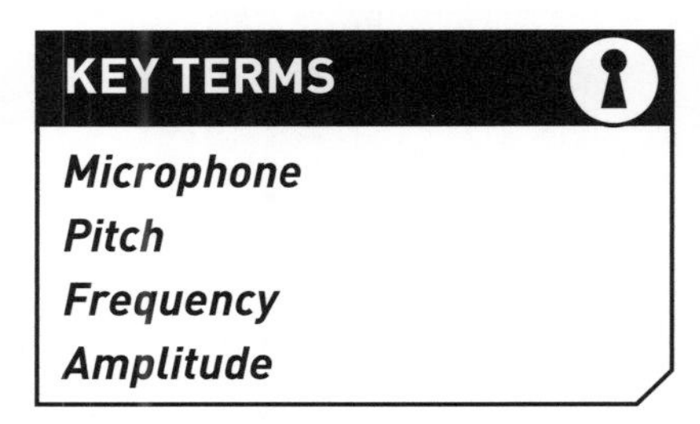

Apparatus

- Cathode ray oscilloscope
- Two tuning forks of different frequency
- Signal generator connected to a loudspeaker
- Microphone

Method

Although your teacher may demonstrate the practical work with the cathode ray oscilloscope (CRO), you should still read through the method to ensure you understand what is happening. You will be asked to record and explain your observations.

SAFETY GUIDANCE 

Keep signal generator frequencies below 20 kHz.

TIP 

A sound wave is converted by the microphone into an alternating voltage; the display on the CRO shows how this voltage varies with time. (A CRO is basically a voltmeter.)

TIP 

The time base setting controls how many waves fit across the screen; the Y-amp gain changes the scale for the p.d. measurement.

1. Switch on the CRO and adjust the focus and brightness of the trace to be sharp and of low intensity.
2. Adjust the time base control so that a horizontal line is obtained on the screen of the CRO.
3. Adjust the Y-shift control so that the horizontal line is in the centre of the screen.
4. Attach a microphone to the Y-input and switch it on.
5. Sound a tuning fork close to the microphone and adjust the time base setting until two or three complete waves are seen on the screen; you may need to change the Y-amp gain so that both the peaks and troughs of the wave are displayed on the screen.
6. Sketch the appearance of the wave.
7. Repeat steps **5** and **6** using a tuning fork of different pitch. Do not alter the time base setting or the Y-amp gain.
8. Sketch the appearance of the wave from the second tuning fork.
9. Investigate the effect of loudness on the amplitude of the waves generated by the tuning forks.
10. Use a signal generator instead of the tuning forks to produce a continuous signal.
11. On the same grid, sketch the appearance of waves with different pitch but the same loudness; label the one that has the highest pitch.
12. On the same grid, sketch the appearance of waves with the same pitch but different loudness; mark the wave corresponding to the loudest sound.

Observations

First tuning fork

Sketch the waveform to scale.

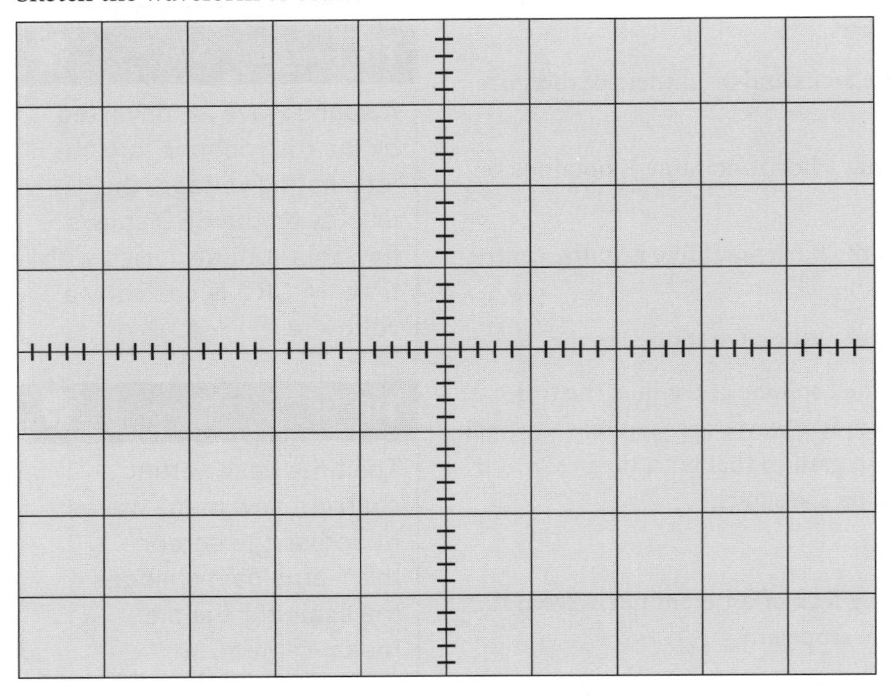

Second tuning fork

Sketch the waveform to scale.

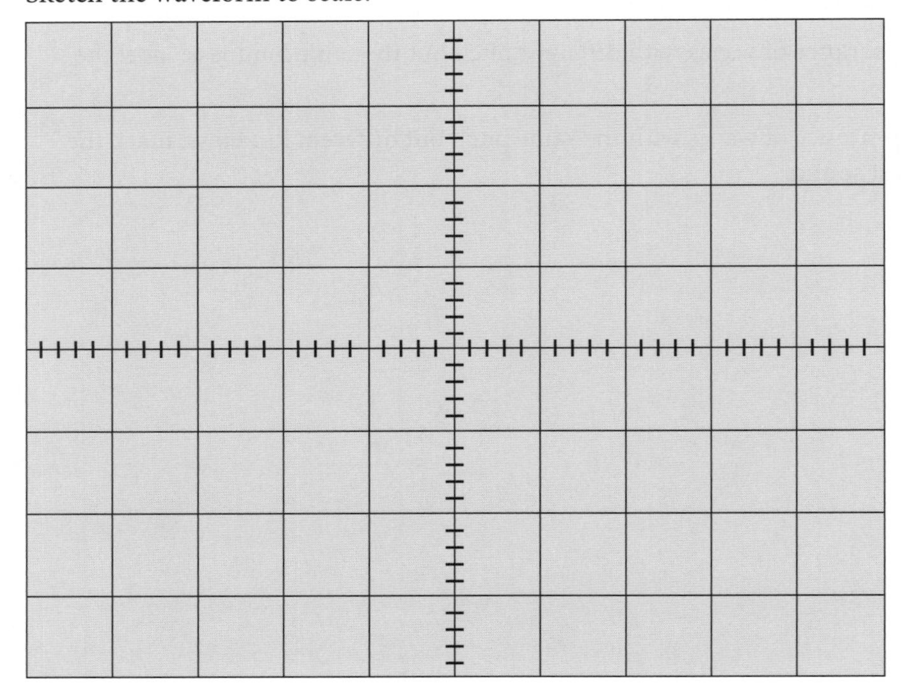

Signal generator: different pitch

Sketch the waveforms to scale.

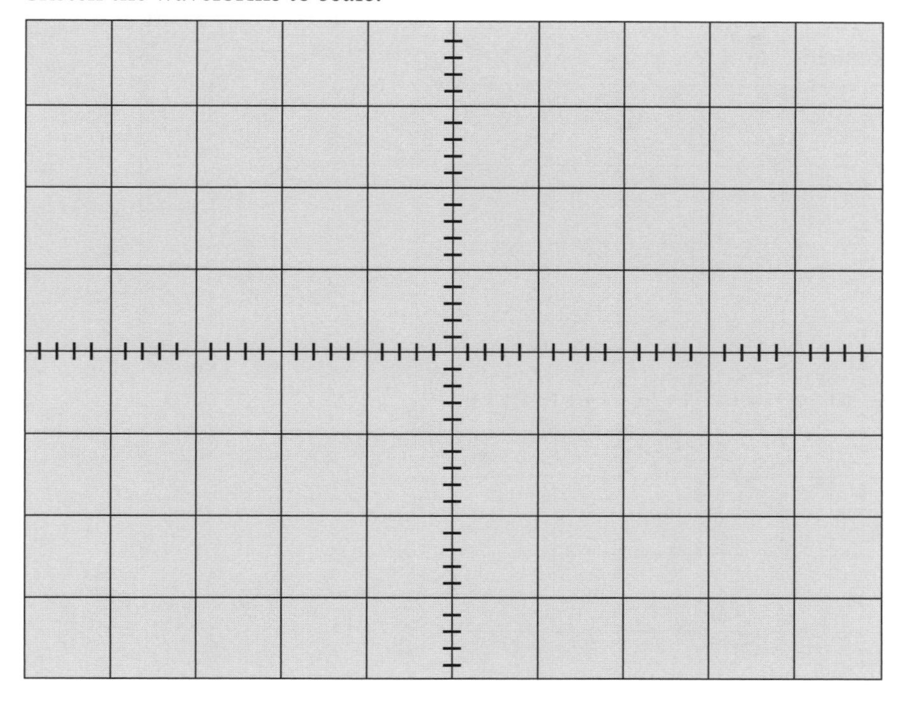

Signal generator: different loudness

Sketch the waveforms to scale.

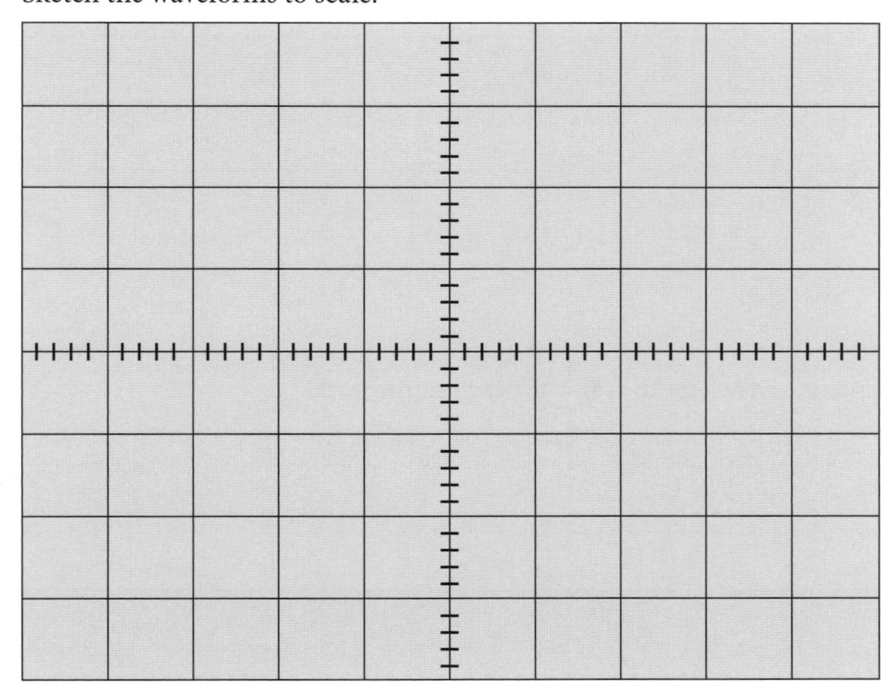

Conclusions

1 On the same grid, sketch and label the waveforms you would expect to see on the CRO for a soft, low frequency note and a loud, high frequency note.

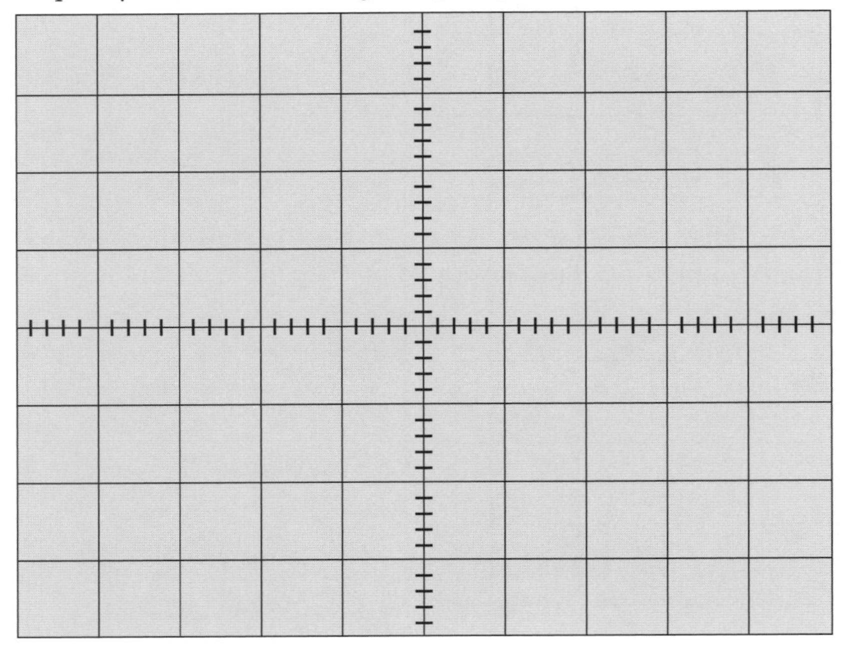

2 Summarise the effects of loudness and pitch on the amplitude and frequency of a sound.

....

....

....

....

Evaluation

Explain why the settings on the oscilloscope were kept the same for each tuning fork.

....

....

GOING FURTHER

Repeat the experiment for one of the tuning forks and estimate its frequency.

You will need to know the time base setting on the CRO.

TIP

If the time base is set to 2 ms/cm, this means that each cm in the horizontal direction of the display represents a time of 2 ms.

4 Electricity and magnetism

4.1 Magnetism

Magnets have two types of magnetic pole: north (N) and south (S). Unlike magnetic poles (N, S) attract, but like magnetic poles (N, N or S, S) repel. A bar magnet has a north pole at one end and a south pole at the other, and this produces a magnetic field around the magnet. A magnetic field can be represented by lines of force. The direction in which the north pole of a compass points gives the direction of the line of force at that point. Iron filings can be used to make magnetic field lines visible.

Aim

To investigate the magnetic field lines around a bar magnet and between two magnetic poles.

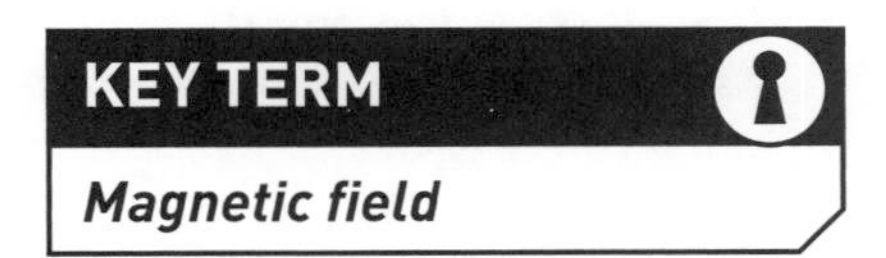

Apparatus

- Eye protection
- Two bar magnets
- Plotting compass
- Iron filings in a shaker
- Paper

Method

Field lines around a bar magnet

1. Lay a bar magnet in the centre of a piece of paper and draw its outline.
2. Place a plotting compass near the north pole of the bar magnet (see point A in Figure 1).
3. Mark the positions of the south and north poles of the compass needle (A, B) on the paper.
4. Move the compass so that the south pole of the compass needle is at the point (B) where the north pole was previously and mark the new position of the north pole of the compass needle (C).

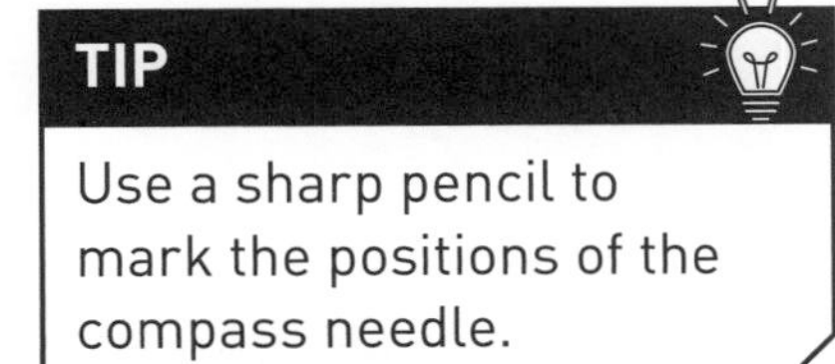

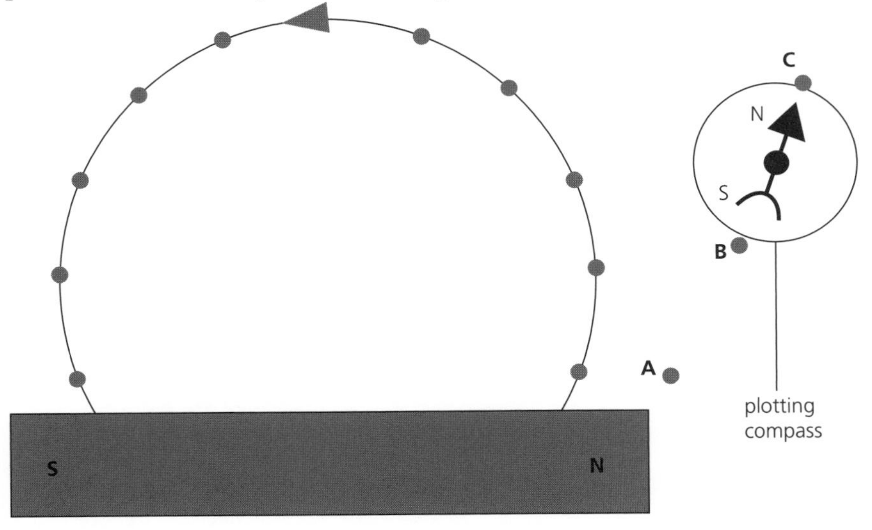

Figure 1

5 Continue step **4** until the compass is near the south pole of the bar magnet.

6 Join up the points to give a field line; mark the direction of the magnetic field with an arrow on the line.

7 Plot other field lines by repeating the steps **2** to **6** with the compass at different starting points.

Field lines between magnetic poles

1 Align two magnets with two like poles opposite each other, about 5 cm apart.

2 Place a piece of paper over the magnets.

3 Sprinkle some iron filings over the paper evenly and thinly; tap the paper gently so that the iron filings line up along magnetic field lines.

4 Sketch the appearance of the field lines below.

5 Repeat steps **1** to **4** with two unlike poles facing each other.

SAFETY GUIDANCE 

Eye protection must be worn when handling and using iron filings.

TIP 

Do not allow iron filings to stick to the magnet.

Observations

Field lines around a bar magnet

Draw a sketch in the space below of the field lines you plotted around the bar magnet.

Field lines between magnetic poles

1 Sketch the field lines between two like poles.

2 Sketch the field lines between two unlike poles.

Conclusions

Complete the following sentences.

1 The magnetic field around a bar magnet is .... near the ends/poles, as is shown by the .... of magnetic field lines in this region.

2 Magnetic field lines are directed from the .... pole of a magnet to the .... pole.

3 Explain why the iron filings line up along magnetic field lines.

....

....

Evaluation

Describe any problems that you had with the experiment. Suggest how these problems could be avoided to produce more reliable results.

....

....

....

....

....

GOING FURTHER

The magnetic tape on the back of a swipe card contains tiny particles made of magnetic material which can be magnetised easily to act as bar magnets with a N pole and a S pole.

1 Suggest how information can be stored on the magnetic tape.

....

....

2 Explain why a swipe card can fail to work after it is placed next to a mobile phone or camera.

....

....

4.2 Electric charges (Teacher demonstration)

There are two types of electric charge, positive and negative. When a material is rubbed with a cloth, electrons (which carry a negative charge) may be transferred by friction to or from the material. If the material *gains* electrons in the process, it becomes negatively charged; a different material may *lose* electrons when it is rubbed and become positively charged. Two negative charges (or two positive charges) repel each other.
A negative and positive charge attract each other. A gold-leaf electroscope can be used to detect charge.

Aim

To investigate the production of electrostatic charges and the forces between electric charges, and to use a gold-leaf electroscope to detect charge and distinguish between good and bad electrical conductors.

KEY TERMS

Charge
Electron
Electrical conductor

Apparatus

- Gold-leaf electroscope
- Polythene strips or rods
- Cellulose acetate strips or rods
- Cloth
- Thread
- Paper stirrup
- Retort stand and clamp
- Some electrically conducting and insulating materials such as plastic, metal and wood

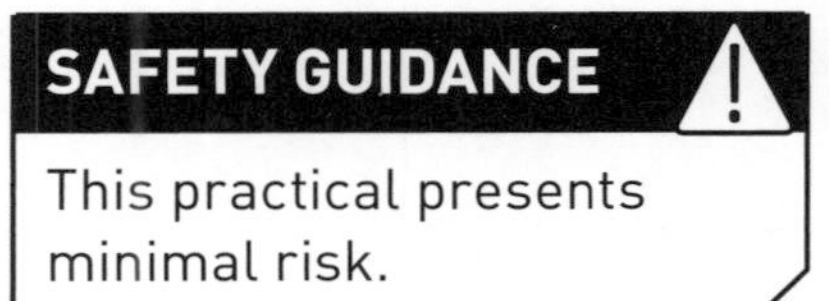

SAFETY GUIDANCE

This practical presents minimal risk.

Method

Positive and negative charges

1. Rub a polythene strip with a cloth.
2. Support the strip horizontally in a paper stirrup suspended from a retort stand; the strip should be able to swing freely.
3. Rub a second polythene strip and bring it close to the suspended strip; record whether the strips are attracted or repelled in Table 1.
4. Rub a strip of cellulose acetate with the cloth and bring it close to the suspended strip; record whether the strips are attracted or repelled.

TIP

Make sure the suspended strip is stationary before bringing up the charged strip.

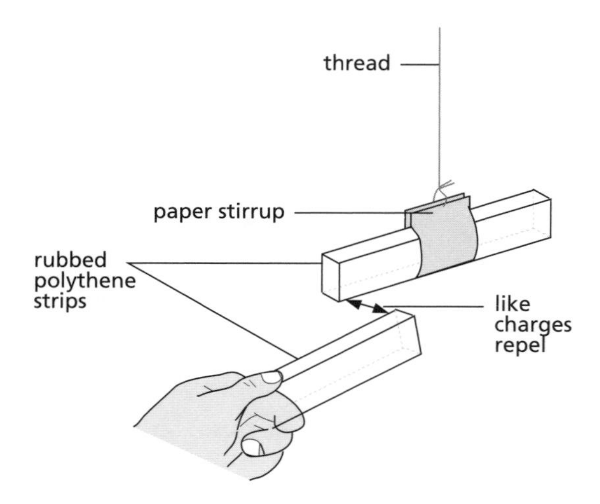

Figure 1

Gold-leaf electroscope

Although your teacher may demonstrate the practical work using the gold-leaf electroscope, you should still read through the method to ensure you understand what is happening. You will be asked to record and explain your observations.

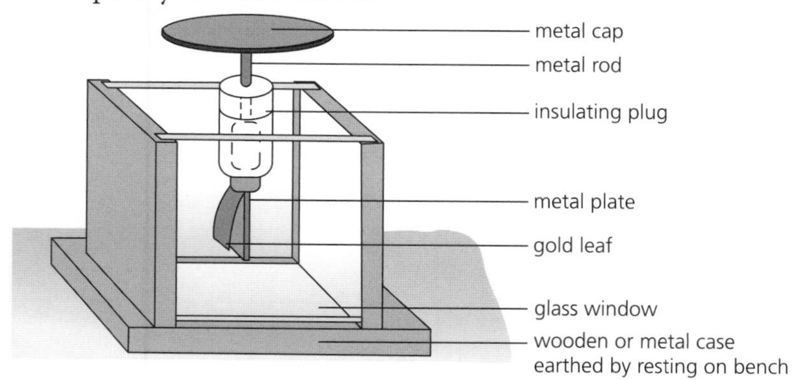

Figure 2

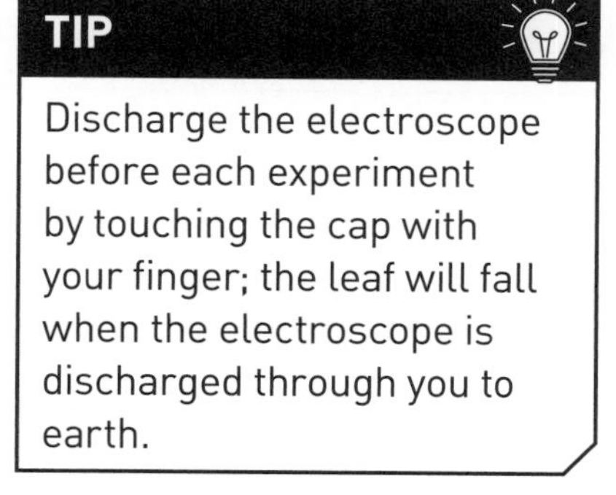

In the Observations section, record what happens to the gold leaf of the electroscope when:

a a charged polythene strip is brought close to (but does not touch) the cap of the electroscope
b the charged polythene strip is moved away from the electroscope
c a charged cellulose acetate strip is brought close to the cap of the electroscope
d the charged cellulose acetate strip is moved away from the electroscope
e a charged polythene strip is drawn firmly across the edge of the cap of the electroscope
f you touch the cap of the charged electroscope with your finger.

To distinguish between good and bad electrical conductors, charge the electroscope as in step **e**. Touch the cap of the charged electroscope with different materials such as plastic, metal and wood; if the gold leaf falls *quickly*, the material is a good conductor. Record your observations in Table 2.

Observations

Positive and negative charges

Complete Table 1.

Table 1

Material of charged strip	Material of charged strip	Attracted or repelled

Gold-leaf electroscope

Complete the following sentences and then complete Table 2 with your observations.

a When a charged polythene strip is brought near the electroscope cap, the gold leaf ....

b When the charged polythene strip is then moved away from the electroscope cap, the gold leaf ....

c When a charged cellulose acetate strip is brought near the electroscope cap, the gold leaf ....

d When the charged acetate strip is then moved away from the electroscope cap, the gold leaf ....

e When a charged polythene strip is drawn across the edge of the electroscope cap, the gold leaf

.... and the electroscope becomes charged.

f When you touch the charged electroscope cap with a finger, the gold leaf .... and the electroscope is discharged through you to earth.

Table 2

Material	Gold leaf falls quickly/slowly	Good/bad conductor

Conclusions

1 Complete the following sentences.

Like charges .... Unlike charges ....

2 Explain what happens when you touch the cap of a charged gold-leaf electroscope with your hand.

....

....

....

3 Explain, in terms of the movement of electrons, what happens when a charged strip is brought close to and then moved away from a discharged gold-leaf electroscope.

....

....

....

....

....

....

....

4 Explain why charging the cap of the electroscope positively or negatively causes the gold leaf to rise.

....

....

....

....

5 Explain how a gold-leaf electroscope can be used

a to detect charge

....

....

....

....

b to distinguish between electrical conductors and insulators.

....

....

....

....

6 List the materials which you found to be

a good conductors: ....

....

b bad conductors: ....

....

Evaluation

Describe any problems that occurred in the experiments and suggest any improvements that could be made.

....

....

....

....

....

....

....

....

....

GOING FURTHER

When a negatively charged polythene rod is brought close to (but not touching) small pieces of aluminium foil, they are attracted to the rod. Can you suggest why?

....

....

....

....

....

....

4.3 Measuring current

An electric current is caused by the flow of electric charges. The unit of current is the ampere (A) and the size of a current is measured by an ammeter. A current transfers energy to devices in a circuit. A current will only flow if there is a complete circuit. In this practical work you will learn how to measure current and connect circuits with devices arranged in series and parallel. Look back at page 9 to remind yourself of how to take an accurate reading from an analogue scale.

Aim

To measure currents in series and parallel circuits.

Apparatus

- Two 1.5 V cells
- Two 1.25 V lamps
- 0–1 A ammeter
- Switch
- Circuit board
- Connecting wire

KEY TERMS

Ammeter
Electric current
Range
Series circuit
Parallel circuit

Method

Series circuits

1 a Set up the circuit shown in Figure 1, making sure that the + terminal of the cell goes to the + terminal (red) of the ammeter.

b Close the switch and record the reading of the current through the ammeter in Table 1.

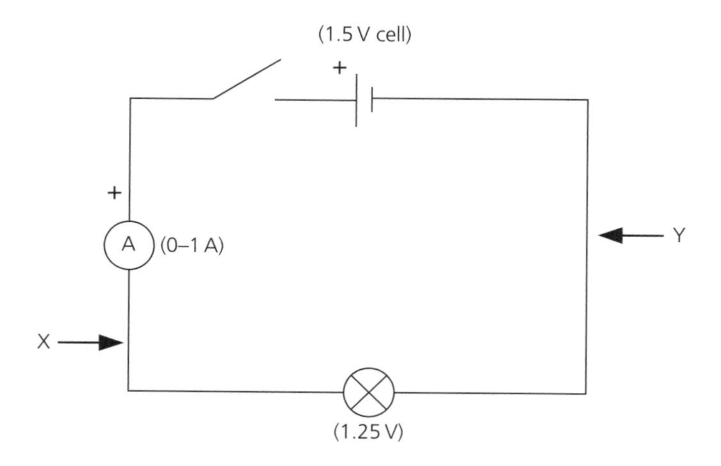

Figure 1

SAFETY GUIDANCE

Switch off power before making changes to a circuit. Large currents can result in burns and electric shocks and can damage sensitive meters.

TIPS

If you are using a meter with a range of settings, select the largest current range setting initially.

Check the components in the circuit are connected correctly before the power is switched on.

TIPS

Make sure the lamp is on when the current reading is being taken.

If the pointer on an analogue meter goes backwards (or the reading is negative on a digital meter), interchange the connections to the ammeter.

If you are using an ammeter with several ranges and the current reading is low, decrease the current range setting to obtain a more precise measurement.

c Disconnect the ammeter and reinsert it in the circuit at position Y. When the circuit is complete, record the current through the ammeter in Table 1.

2 a Connect two cells and two lamps *in series* with the ammeter as shown in Figure 2. The + terminal of one cell should be connected to the – terminal of the other. When the switch is closed and both lamps are on, record the current reading through the ammeter at A in Table 1.

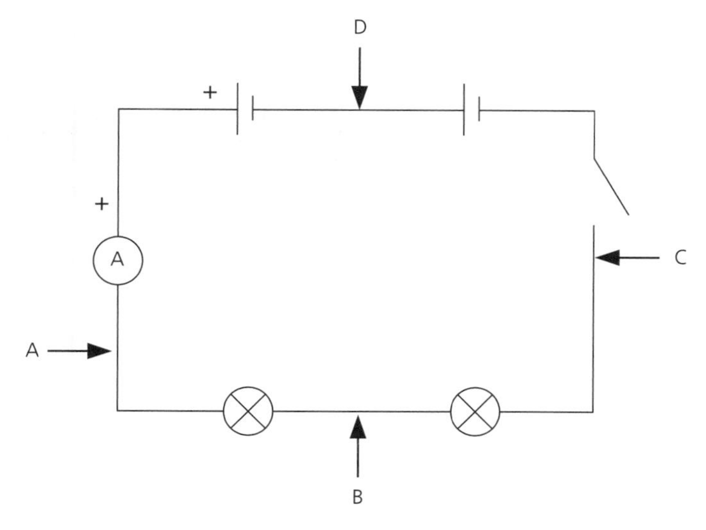

Figure 2

b Disconnect the ammeter and reinsert it in the circuit at position B. When the circuit is complete, record the value of the current through the ammeter in Table 1.

c Repeat step **2b** with the ammeter inserted into positions C and then D in the circuit.

Parallel circuits

3 a Connect the two lamps *in parallel* with one cell as shown in Figure 3. When the switch is closed and both lamps are on, record the current passing through the ammeter at S in Table 1.

b Disconnect the ammeter and reinsert it in the circuit next to the lamp at position P. When the circuit is complete and both lamps are on, record the value of the current through the ammeter at P in Table 1.

c Repeat step **3b** with the ammeter inserted into positions Q and then R in the circuit.

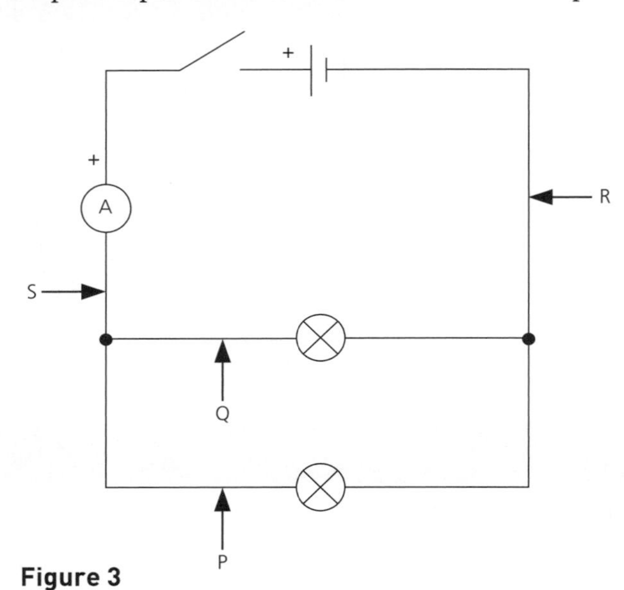

Figure 3

Observations

Complete Table 1.

Table 1

Ammeter position	X	Y	A	B	C	D	S	P	Q	R
Current/A										

Conclusions

Complete the following sentences.

In a series circuit, the current is ....

at all points within experimental error.

In a parallel circuit, the current from the cell is .... than

the current in each branch.

The sum of the currents in the branches of a parallel circuit equals the

current .... or leaving the junction of the

parallel section within experimental error.

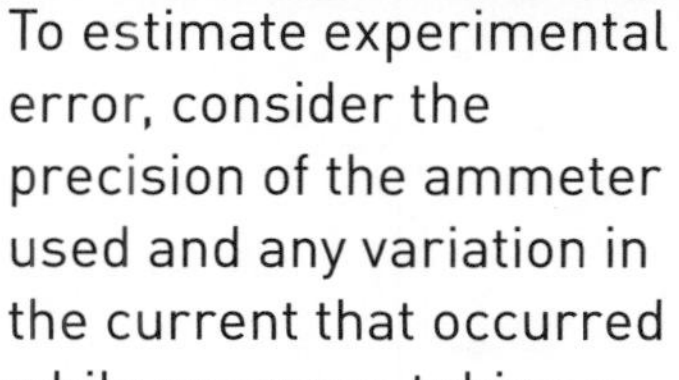

TIP

To estimate experimental error, consider the precision of the ammeter used and any variation in the current that occurred while you were taking a measurement.

Evaluation

1 Describe any difficulties you encountered and improvements that could be made in performing the experiments.

....

....

....

....

....

....

....

2 You are provided with a 1.5 V cell, a 1.25 V lamp, a switch, some crocodile clips and connecting wires. Describe the circuit you could set up to test whether a sample of material is a good or a bad electrical conductor and explain how it works.

....

....

....

....

....

....

....

4.4 Measuring resistance

Resistance is the opposition of a conductor to the flow of an electric current and is measured in ohms (Ω). The resistance of a conductor can be determined by measuring the potential difference (p.d.) across the conductor with a voltmeter, and the current (I) flowing through it with an ammeter. A good electrical conductor has a low resistance and a poor electrical conductor has a high resistance. The resistance of a wire depends on the material it is made of and its physical dimensions.

Aim

To measure the resistance of a wire and investigate how the resistance depends on the length and diameter of the wire.

KEY TERMS

Resistance
Potential difference (p.d.)
Voltmeter
Variable resistor
Ohm

Apparatus

- 4.5 V battery (three 1.5 V cells connected in series can be used)
- Ammeter (0–1 A)
- Voltmeter (0–5 V)
- 0–25 Ω rheostat (variable resistor)
- Constantan wire (two different diameters are required)
- Metre ruler
- Wires and connectors
- Crocodile clips
- Insulated circuit board with mounting clamps

KEY EQUATION

The resistance, R, of a component is given in ohms by:

$$R = \frac{V}{I}$$

where V is the p.d. across the component in volts and I is the current flowing through it in amperes.

Method

Measurement of resistance

Before you take any measurements, read the instructions below. Complete Table 1 to identify which variable you will change, which variables you will keep fixed and which variable you will measure.

Table 1

Independent variable (variable you change)	Control variables (constant/fixed)	Dependent variable (variable you measure)

SAFETY GUIDANCE

Turn off the current between readings for safety and to try to keep the temperature of the wire constant.
The constantan wire may become hot with larger currents.

1. Set up the circuit shown in Figure 1 where the unknown resistance, R, is a 1 m length of constantan wire held above the circuit board in two vertical clamps.
2. Set the rheostat at its maximum resistance.
3. Record the current, I, in the circuit and the p.d., V, across the unknown resistance, R, in Table 2.
4. Reduce the resistance of the rheostat and record the new values of I and V.
5. Repeat step **4** until you have six different readings of I and V. Calculate R for each pair of readings.
6. Plot a graph of V along the x-axis and I along the y-axis.

TIP

The variable resistor allows the current to be changed in the wire. Carefully check that you have connected the components correctly before completing the circuit.

TIPS

Choose an easy-to-read/plot scale and use as much of the graph paper as possible.

Remember to give units when you are labelling the axes and plot the data points clearly with a sharp pencil.

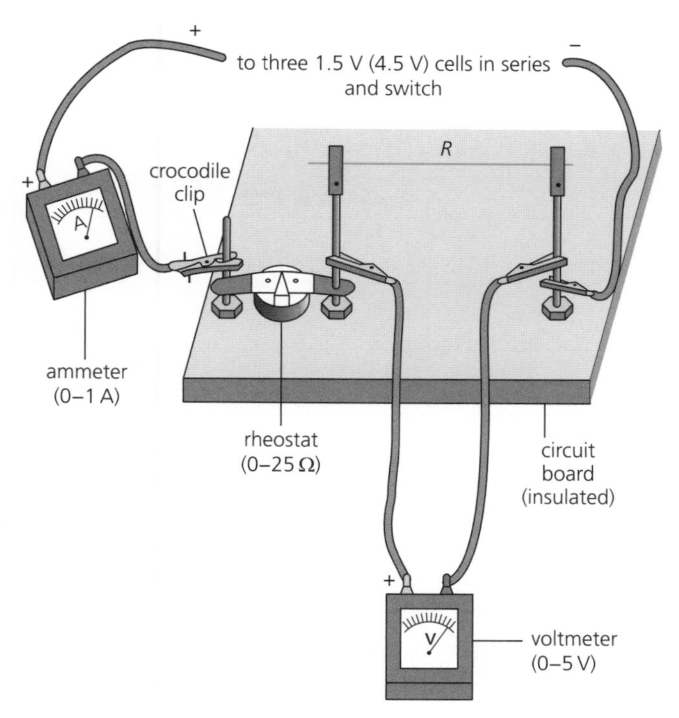

Figure 1

Effect on resistance of the length and diameter of the wire

1 Record in Table 3 a pair of mid-range values for *I* and *V* read from your *I*–*V* graph.

Replace the 100 cm long wire used above by a constantan wire of the same thickness but of length 50 cm.

2 Record the value of the diameter of the wire in Table 3.

3 Set the rheostat at mid-range, switch on and record the current in the circuit and the p.d. across the wire in Table 3. *Switch off the power.*

4 Replace the wire used in step **3** with a constantan wire of a different diameter, again 1 m long. Record the value of the diameter of the wire in Table 3. Without changing the setting of the rheostat, switch on the power and record the current in the circuit and the p.d. across the wire. *Switch off the power.*

5 Repeat step **4** with a 50 cm length of the same diameter constantan wire that was used in step **4**.

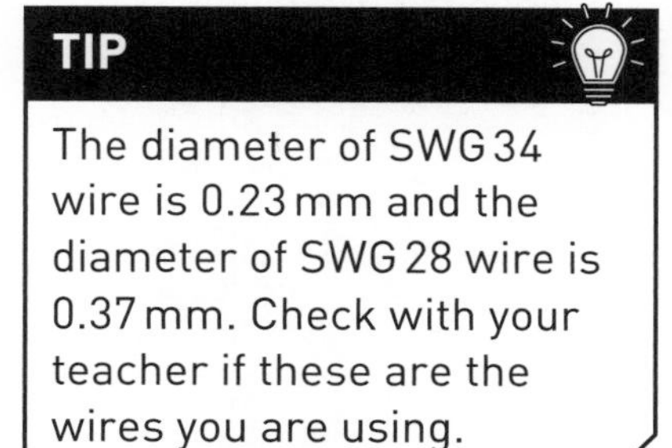
TIP

The diameter of SWG 34 wire is 0.23 mm and the diameter of SWG 28 wire is 0.37 mm. Check with your teacher if these are the wires you are using.

Observations

Draw a circuit diagram of the experimental arrangement.

Measurement of resistance

1 Complete Table 2.

Table 2 Calculation of resistance

Current, I/A	p.d., V/V	Resistance, R/Ω

2 Plot a current–voltage graph from your results. (See point 6 and the Tips box on page 96.)

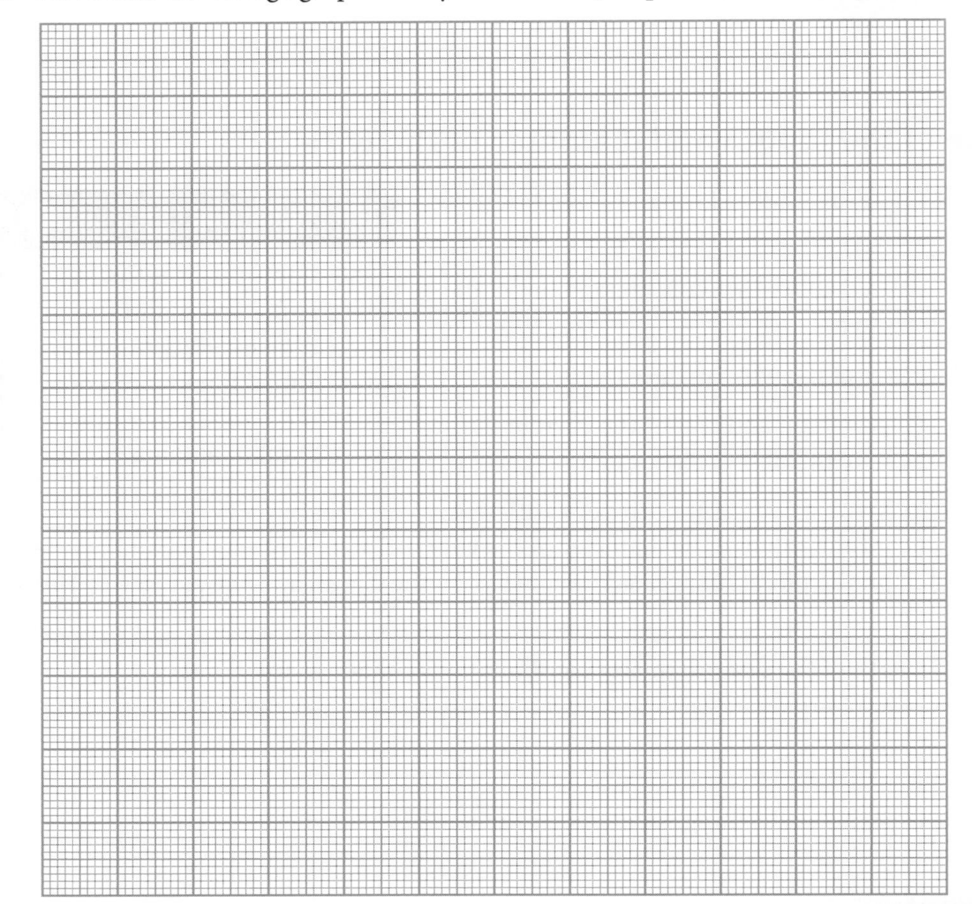

3 Find the gradient of your graph.

Gradient of graph = ... A/V

$\frac{1}{\text{gradient}}$ of graph = ... Ω

TIP

Choose the best straight line through your points, use the triangle method over as long a length of line as possible and show your working.

Effect on resistance of the length and diameter of wire

Complete Table 3.

Table 3 Variation of resistance with length and diameter of a wire

Length of constantan wire/cm	Diameter of wire/mm	Current, I/A	p.d., V/V	Resistance, R/Ω
100				
50				
100				
50				

TIP

Use a pair of readings from your I–V curve for the first row of Table 3.

Conclusions

1 **a** State how the resistance of the wire changes when the current through it increases. Use data from Table 2 to support your answer.

....

....

b With reference to your I–V graph, state how the current through the constantan wire is related to the p.d. applied across it.

....

....

2 State how the resistance of the wire varies with the length and diameter of the wire. Use data from Table 3 to support your answer.

....

....

....

....

....

Evaluation

1 Suggest how you would expect your I–V plot to change if the resistance of the constantan wire changed with temperature.

....

....

....

2 Discuss how the experiment could be improved to increase the accuracy of the calculated resistance or to increase the validity of the conclusions.

....

....

....

....

GOING FURTHER

A student suggests that the resistance of a wire is inversely proportional to its cross-sectional area.

Describe how you could investigate the relationship between the resistance, length and diameter of a wire, and use the data to test the student's hypothesis.

....

....

....

....

....

....

....

4.5 Potential divider

A potential divider consists of two resistors connected in series with a fixed supply voltage.

If one of the resistors is a variable resistor and its value is changed, the p.d. increases across one resistor and decreases across the other. In this practical you will use a voltmeter to measure the p.d. across each resistor when the resistance of the variable resistor is changed.

Thermistors and light-dependent resistors are often used as the variable resistor in a potential divider circuit designed to monitor temperature or light intensity.

KEY TERMS 

Resistance
Potential difference
Thermistor
Potential divider

KEY EQUATION 

In a potential divider, the ratio of the voltages V_1 and V_2 across resistances R_1 and R_2 is given by:

$$\frac{V_1}{V_2} = \frac{R_1}{R_2}$$

Aim

To investigate the action of a potential divider.

Apparatus

- 6 V battery or d.c. power supply
- Ammeters (0–50 mA and 0–1 mA)
- Two voltmeters (0–10 V)
- 150 Ω and 10 kΩ fixed resistors
- 150 Ω variable resistor
- Thermistor (TH7)
- Matches
- Switch
- Wires and connectors

Method

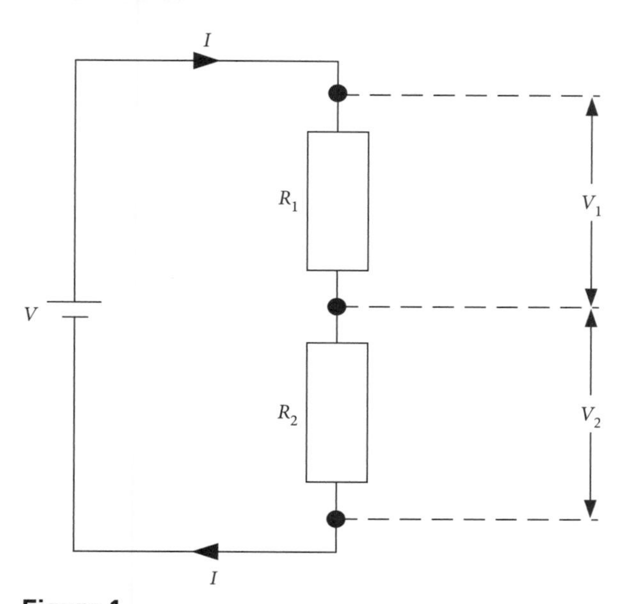

Figure 1

SAFETY GUIDANCE 

- To reduce risk of electric shock, do not set up electric circuits near water (sink or tap).
- Switch off supply voltage before changing components.
- Take care not to burn your fingers when using the matches.

Variable resistor

1 Set up the circuit shown in Figure 1 with R_1 as the 150 Ω fixed resistor and R_2 as a variable 150 Ω resistor.

2 Insert a switch and ammeter (0–50 mA) into the circuit to measure the current, I.

3 Connect a voltmeter across each of R_1 and R_2 to measure V_1 and V_2.

4 Switch on the power supply (6.0 V) and record the values of I, V_1 and V_2 in Table 1.

5 Change the value of the variable resistor until V_1 = 3.5 V; record the new values of I, V_1 and V_2 in Table 1.

6 Repeat step **5** for successive values of V_1 of 4.0 V, 4.5 V and 5 V.

TIP 

Set the variable resistor to its maximum value.

TIP 

The ammeter should be connected in series with the resistors (see page 93).

Thermistor

1 Repeat steps **1** to **3** of Experiment 1 but with R_1 as the 10 kΩ fixed resistor and R_2 as a thermistor.

2 Switch on the power supply and record the values of I, V_1 and V_2 in Table 2.

3 Warm the thermistor with a match and record the new values of I, V_1 and V_2.

4 As the thermistor cools, record a third set of values for I, V_1 and V_2.

Observations

Variable resistor

Complete Table 1.

Table 1

I/mA	V_1/V	V_2/V	$\frac{V_1}{V_2}$	R_2/Ω	$\frac{R_1}{R_2}$

Thermistor

Complete Table 2.

Table 2

I/mA	V_1/V	V_2/V	$\frac{V_1}{V_2}$	R_2/Ω	$\frac{R_1}{R_2}$

Conclusions

1 With reference to your results from the two experiments, state how the p.d.s across the resistors R_1 and R_2 change when the value of the variable resistor R_2 is

a decreased

....

b increased.

....

2 Explain how you could use the potential divider circuit to obtain voltages in the ratio of 3 : 1 from the fixed 6 V supply.

....

....

....

3 State whether your results are in agreement with the potential divider equation.

....

4 When the temperature of the thermistor rises, state whether an increase or decrease occurs in:

a I ....

b V_1 ....

c V_2 ....

d R_2 ....

5 Explain how the potential divider circuit with a thermistor could be used to monitor temperature.

....

....

....

....

Evaluation

1 The potential divider circuit you built has a battery with a fixed e.m.f., one fixed resistor and one variable resistor.

Give a reason why being able to change the value of the variable resistance makes this circuit useful.

....

....

2 Describe any difficulties you encountered in the experiment and how you overcame them.

....

....

....

3 Suggest any improvements that could be made to the experiment.

....

....

....

GOING FURTHER

Explain how a thermistor in a potential divider circuit with a relay coil as the second resistor could be used as a high-temperature alarm.

....

....

....

....

....

Draw a circuit diagram that could be used for this circuit; include a thermistor, resistor, relay and bell.

4.6 Measuring electrical power

An electric motor transfers electrical energy into motion. Energy is wasted in the transfer due to electrical heating, and due to friction in moving parts. The ratio of the output power to the input power gives a measure of the efficiency of the motor.

Aims

To measure the input and output power of an electric motor.

To measure the efficiency of an electric motor.

KEY TERMS 

Gravitational potential energy
Work
Efficiency
Power
Gravitational field strength

KEY EQUATIONS

Power, $P = \frac{\text{energy transferred or work done}}{\text{time taken}}$

$P = \frac{\Delta E}{t}$ or $P = \frac{W}{t}$

The rate at which electrical energy is transferred to a device, P, is given by:

$P = IV$

where I is the current through the device and V is the p.d. across it.

The change in gravitational potential energy when a mass, m, is lifted through a height, Δh, through a gravitational field of strength g is given by:

$\Delta E_p = mg\Delta h$

The efficiency of a device is defined by:

(%) efficiency = $\frac{\text{(useful power output)}}{\text{(total power input)}} \times 100\%$

Apparatus

- 9 V battery
- Ammeter (0–2 A)
- Voltmeter (0–10 V)
- Small electric motor
- Switch
- String
- 100 g masses
- Timer
- Metre ruler
- Clamp
- Wires and connectors

Method

1 Connect the battery, ammeter and electric motor in series as shown in Figure 1 overleaf.

2 Connect the voltmeter across the terminals of the electric motor.

3 Attach a 100 g mass to the axle of the motor with a piece of string and allow the mass to hang freely over the edge of the bench (Figure 2). Clamp the motor to the bench.

4 Support the metre ruler vertically behind the hanging mass.

5 Start the motor and record in Table 1 the time, t, it takes for the mass, m, to rise smoothly through a height, Δh, of 0.50 m. In Table 2, record the ammeter and voltmeter readings (I and V) while the mass is rising.

6 Increase the load by 100 g and repeat step **5**.

7 Repeat step **6** two more times.

SAFETY GUIDANCE

- Clamp the motor firmly to the bench.
- Protect feet from falling masses.
- Stop the motor before the mass reaches the motor axle.

TIP

It is better to work in groups of three for this experiment as there are several sets of measurements to take at the same time.

TIPS

If the mass sways initially, wait until it is rising smoothly before timing the rise.

Plan where you will start and stop timing to give a rise of 0.50 m.

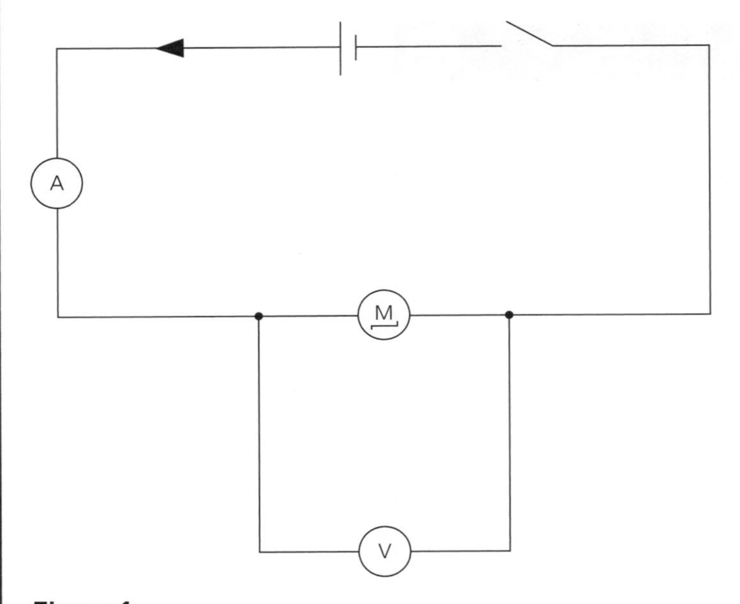

Figure 1

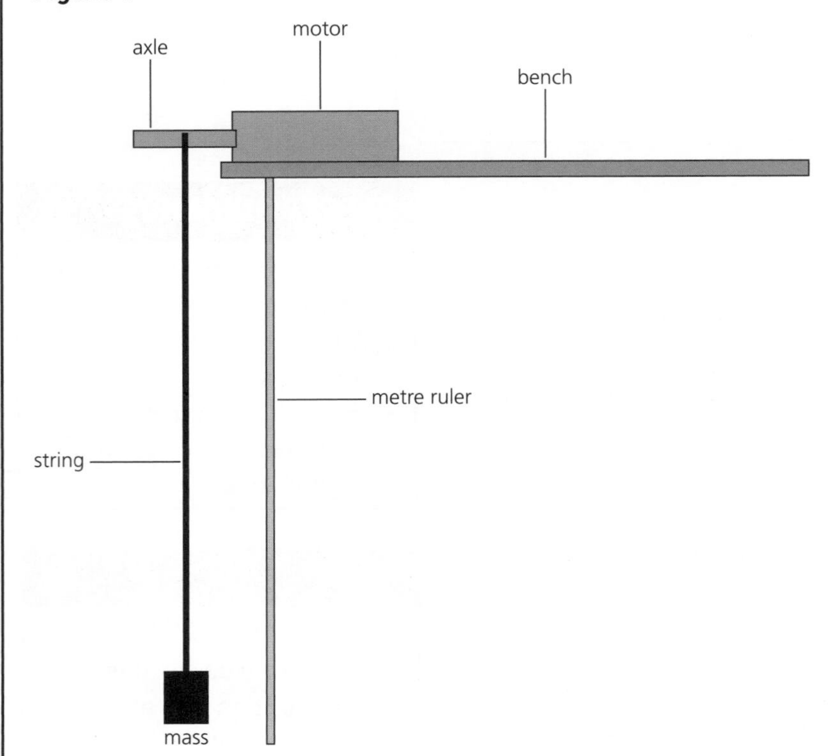

Figure 2

Observations

Record your measurements in Tables 1 and 2.

Calculate and record the input power of the electric motor using $P = IV$.

Calculate and record the work done by the motor in lifting the mass, using $\Delta E_p = mg\Delta h$. Take $g = 9.8$ N/kg.

Calculate and record the output power of the motor using:

$$P = \frac{mg\Delta h}{t}$$

Calculate the efficiency of the motor for different sizes of load.

TIP

Calculate values to two significant figures.

Table 1

Mass raised, m/kg	Time taken to raise mass, t/s	Height mass raised, Δh/m	Gain in potential energy of mass, $mg\Delta h$/J	Rate at which work done in raising mass, $mg\Delta h/t$/W

Table 2

Mass raised, m/kg	Current in motor, I/A	p.d. across motor, V/V	Power input to motor, $P = IV$/W	Useful power output, $mg\Delta h/t$/W	Efficiency of motor %

Conclusions

1 Summarise your results.

....

....

....

....

....

2 Is the efficiency of the motor affected by the size of the load? ....

3 Is the speed of the motor affected by the size of the load? ....

4 Explain why the efficiency of the motor is always less than 100%.

....

....

....

Evaluation

1 Describe at least two precautions you took to ensure accurate measurements of the time taken to raise the mass.

....

....

....

....

2 Suggest how the experiment could be improved and what effect the improvement would have.

....

....

....

....

....

....

....

4.7 Electromagnetism

A magnetic field is produced around a current-carrying wire. You can make an electromagnet by passing a current through a coil of wire. The strength of the electromagnet is increased if it has an iron core that becomes magnetised when the current is switched on. The magnetic field around a solenoid is the same as that around a bar magnet.

An e.m.f. is induced in a conductor when there is relative motion between the conductor and a magnetic field. If the conductor is in a complete circuit, the induced e.m.f. produces a current in the circuit; the larger the induced e.m.f., the larger the current that flows.

Aim

To construct and test an electromagnet, and to investigate the factors affecting the size of the induced e.m.f. when a magnetic field is moved near a conductor.

KEY TERMS

Electromagnet
Solenoid
Magnetic field
Electromotive force (e.m.f.)
Electromagnetic induction

Apparatus

- 3 m of PVC-covered copper wire (SWG26) with bare ends
- 5 cm long iron nail
- Retort stand and clamp
- Switch
- Paper clips
- Ammeter (0–2 A)
- Rheostat (0–15 Ω)
- 3 V battery
- Plotting compass
- Bar magnet
- 600-turn coil
- Sensitive centre-zero meter (microammeter or voltmeter)
- Wires and connectors

Method

Electromagnet

1 Leave about 25 cm at each end of the wire (for connecting to the circuit) and then wind about 50 cm of wire in a single layer onto the nail. Keep the turns close together; always wind in the same direction.

2 Set up the circuit shown in Figure 1 with the rheostat (variable resistor) set at maximum resistance.

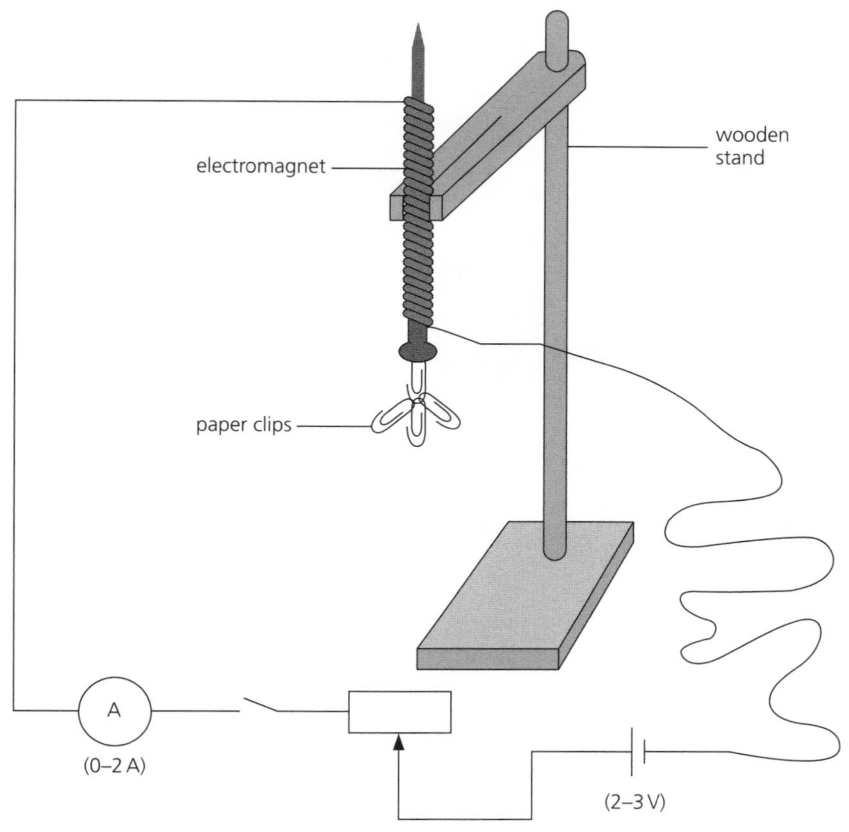

Figure 1

3 In Table 1, record the number of paper clips the electromagnet can support for a range of currents between 0.2 A and 2 A, including 1 A. For the 1 A current, also record the number of paper clips in Table 2.

4 Wind a second layer of wire back along the nail, winding in the same direction as the first layer.

SAFETY GUIDANCE 

The wire becomes hot when current flows through it so take care not to touch it. Only switch on for a few seconds.

TIP 

Keep approximate count of the number of turns as you wind the wire around the nail.

TIPS 

Close the switch only when you are ready to take a reading and switch the current off between readings.

5 In Table 2, record the number of paper clips now supported when the current in the coil is 1 A.

6 Use a plotting compass to determine the N pole of the coil.

7 Reverse the direction of current flow through the coil (by interchanging the connections to the battery) and again determine the position of the N pole.

Electromagnetic induction

1 Connect the 600-turn coil directly to a sensitive centre-zero meter as shown in Figure 2.

2 Identify the N pole of the bar magnet with the plotting compass.

3 In Table 3, record the direction and value of the maximum deflection of the meter when the N pole of the bar magnet is moved *quickly towards* the coil along the line of its axis.

4 Record the direction and value of the maximum deflection of the meter when the N pole of the bar magnet is moved *slowly towards* the coil.

5 Repeat steps **3** and **4** with the N pole of the magnet moving *away* from the coil.

6 In Table 4, record the direction and value of the maximum deflection of the meter when the coil is moved *quickly towards* the N pole of the magnet.

7 Record the direction and value of the maximum deflection of the meter when the coil is moved *quickly away from* the N pole of the magnet.

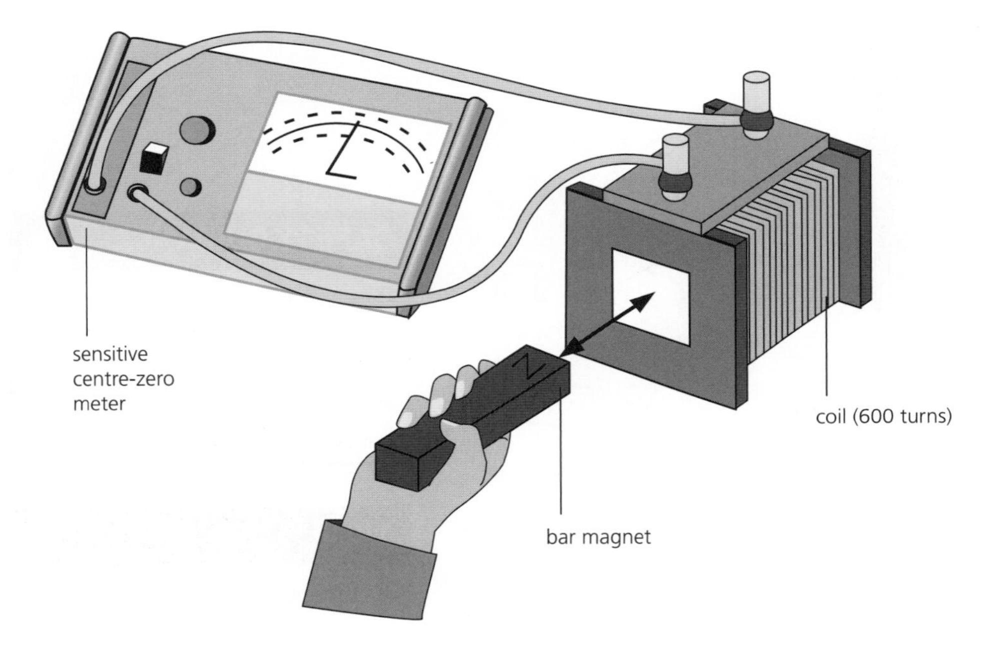

Figure 2

Observations

Electromagnet

1 Complete Tables 1 and 2.

Table 1

Current, *I*/A	Number of paper clips supported

Table 2 Current in coils = 1 A

Number of turns on coil (approx.)	Number of paper clips supported

TIP

Check your answer using the right-hand-grip rule which states that if the fingers of the right hand grip the coil of an electromagnet in the direction of the current flow, the thumb points to the N pole of the coil.

2 The top of the coil is a .... pole when current flows down the coil.

3 State the effect of reversing the current in the coil.

....

Electromagnetic induction

Answer the following question and complete Tables 3 and 4.

What does a change in the direction of the deflection of the needle in the meter indicate?

....

....

....

Table 3 Movement of magnet

Direction of movement of N pole	Speed of movement of magnet	Maximum meter reading	Direction of deflection of needle on meter

Table 4 Movement of coil

Direction of movement of coil	Speed of movement of coil	Maximum meter reading	Direction of deflection of needle on meter

Conclusions

1 State how the strength of the electromagnet depends on the current and the number of turns on the coil.

....

....

....

....

2 What happens to the needle on the meter when the magnet stops moving?

....

....

3 State how an e.m.f. can be induced in a coil.

....

....

....

4 How can the size of an induced e.m.f. be increased?

....

....

....

Evaluation

Suggest how the experiments could be improved to give more reliable results.

....

....

....

GOING FURTHER

If a magnet is moved into or out of a coil, the current induced in the coil makes it into an electromagnet with a N pole and a S pole.

If the N pole of a bar magnet moves towards a coil, will the end of the coil closest to the magnet become a N pole or a S pole? Justify your answer in terms of magnetic forces, work done and the principle of conservation of energy.

....

....

....

....

....

....

4.8 Investigating beams of electrons (Teacher demonstration)

Beams of electrons moving at a high speed are called cathode rays. They can be produced by heating a metal filament in a vacuum and are accelerated by applying a potential difference between an anode and a cathode. When a beam of electrons strikes a fluorescent screen, it causes the screen to fluoresce with a green or blue light enabling the position of the electron beam to be detected.

Charged particles such as electrons are deflected in magnetic and electric fields. This property is exploited in the cathode ray oscilloscope (CRO).

KEY TERMS 

Electron
Cathode rays
Magnetic field
Fleming's left-hand rule

Aim

To study the effect of a magnetic field on a beam of electrons with a Maltese cross tube.

Apparatus

- Eye protection
- Maltese cross tube
- 3 kV power supply
- 6 V power supply for the heater filament
- Bar magnet

SAFETY GUIDANCE 

- A high-voltage power supply is used which can be dangerous.
- There is a risk of the evacuated tube imploding. The demonstrator must wear eye protection and not stand close to the evacuated tube. The demonstrator should either use a safety screen between the apparatus and students, or all students must wear eye protection.

Method

Although your teacher may demonstrate the practical work with the Maltese cross tube, you should still read through the method to ensure you understand what is happening. You will be asked to record and explain your observations.

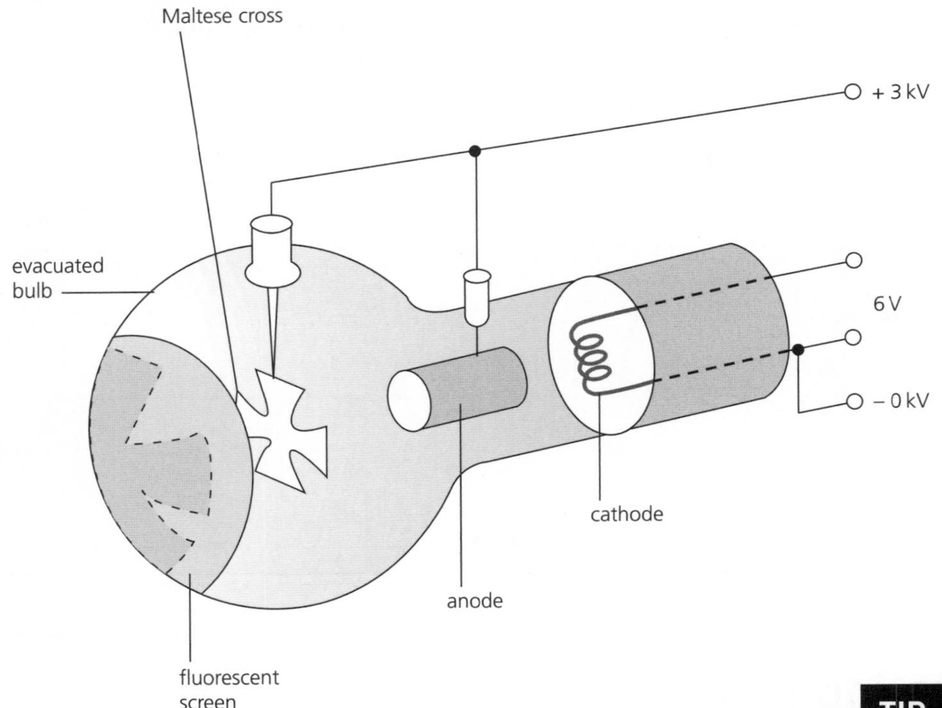

Figure 1

1 Connect the power supplies to the Maltese cross tube according to the manufacturer's instructions and adjust the settings to obtain a shadow of the cross on the fluorescent screen.
2 To observe the deflection of cathode rays in a magnetic field, bring the north pole of a bar magnet close to the neck of the tube. Record the direction in which the rays and the fluorescent shadow move.
3 Now bring the south pole of the magnet close to the neck of the tube and record the result.

TIP

It will help to see the shadow if the lights in the laboratory are dimmed.

TIP

Note that the optical shadow of the cross, due to light emitted by the heated cathode, does not move and is unaffected by the magnetic field.

Observations

1 State how you know that the cathode rays travel in straight lines.

....

....

....

....

2 When the north pole of a magnet is brought close to the neck of the tube the image of the Maltese cross moves ....

3 When the south pole of a magnet is brought close to the neck of the tube the image of the Maltese cross moves ....

Conclusions

1 State the direction of the force on the electron beam when the north pole of the magnet was brought close to the tube.

....

2 State the source of the force on the electron beam.

....

3 Use Fleming's left-hand rule to determine the direction of the current in the Maltese cross and hence the charge on the cathode rays.

....

....

....

....

....

Evaluation

Summarise what you learnt about a beam of electrons in this demonstration.

....

....

....

....

....

....

....

....

5 Space physics

5.1 Phases of the Moon

Some experiments take hours, days, weeks and even years for results to be obtained. To study the phases of the Moon you will need to collect data over several weeks.

For thousands of years, different cultures have used observations of the Sun and Moon to produce a calendar for the year. The modern calendar is based on a year of 365 days, the approximate time it takes for the Earth to orbit the Sun. The Lunar calendar year is shorter, with 12 repeats of the 29.5 days it takes for the Moon to orbit the Earth. This is why the dates of festivals which are traditionally determined from the Lunar calendar occur on different dates each year, for example, Ramadan, Chinese New Year and Easter.

Aim

To record the phases of the Moon over the period of a month.

Apparatus

- Clock or watch
- Compass

SAFETY GUIDANCE 

This practical presents minimal risk.

Method

1. Find out at what time the Moon rises and sets.
2. Look at the Moon and identify its phase (new Moon, crescent Moon, half Moon, full Moon).
3. Record your results in Table 1 and include a sketch of the appearance of the Moon.
4. Repeat your readings about every three days over a period of a month.
5. Record the direction in which the Moon rises and sets.

TIP 

Your local newspaper may give some information on the times of Moon rise and Moon set. Alternatively look it up on the internet using the search term 'moon rise'.

TIP 

A full Moon is visible only between sunset and sunrise but other phases can be seen during daylight hours.

Observations

Direction of Moon rise and set:

The Moon rises in the ....

The Moon sets in the ....

Complete Table 1.

Table 1

Date Moon observed	Phase of Moon	Sketch	Time of Moon rise	Time of Moon set

Conclusions

Complete the following sentences.

1 The Moon rises at a .... time each day.

2 It takes .... days for the phases of the Moon to complete one cycle.

Evaluation

1 Describe any difficulties you encountered in collecting the data.

....

....

....

....

2 Where does the light which makes the Moon visible come from?

....

3 Why is the full disk of the Moon not always visible?

....

....

4 State the phases of the Moon when it is visible in the sky during the day. Use data from Table 1.

....

....

5 Make a sketch to illustrate how the phases of the Moon arise. Include rays from the Sun, the Earth and different positions of the Moon around the Earth. Follow the steps given here:

- Draw a circle approximately 6 cm diameter in the middle of the space below to represent the orbit of the Moon. At the centre of the circle draw another circle with a diameter of about 1 cm and label this 'Earth'.
- Draw part of another circle (not to scale) at the far right side, or below the Earth, to represent the Sun.
- Draw the Moon at eight different positions around the Earth.
- Show parallel rays of light from the Sun towards the Earth.
- Shade in the portion of the Moon that would appear dark, as viewed from Earth.
- Label the part of each Moon circle that represents the visible phase of the Moon.

6 Look at your diagram. Does full Moon occur when the Moon is nearer to or further from the Sun than the Earth? ....

GOING FURTHER

Suggest a reason why the Moon rises at a different time each day.

....

....

....

....

Past paper questions

Practical Test past paper questions

1 In this experiment, you will determine the weight of a metre rule using a balancing method.
Carry out the following instructions, referring to Figure 1.

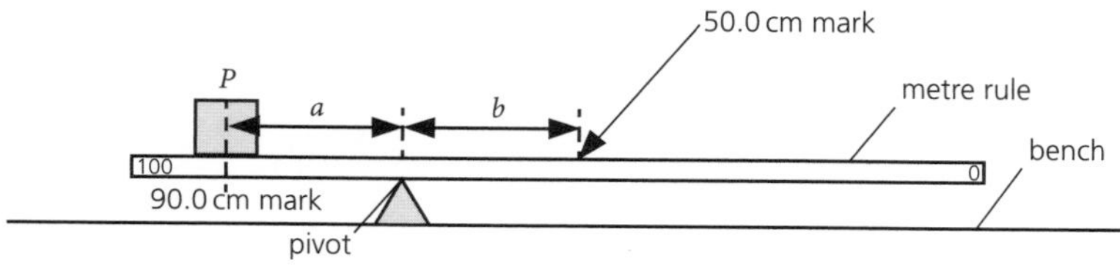

Figure 1

a Place the metre rule on the pivot. Place the load P with its centre on the metre rule at the 90.0 cm mark. Keeping the load P at the 90.0 cm mark, adjust the position of the metre rule on the pivot so that the metre rule is as near as possible to being balanced.
Measure, and record in the first row of Table 1, the distance a from the 90.0 cm mark to the pivot.
Measure, and record in the first row of Table 1, the distance b from the pivot to the 50.0 cm mark.
Repeat the steps above, placing the centre of the load P at the 85.0 cm, 80.0 cm, 75.0 cm and 70.0 cm marks. Record all values of a and b in Table 1. [3]

Table 1

a/cm	b/cm

b Plot a graph of a/cm (y-axis) against b/cm (x-axis). You do **not** need to begin your axes at the origin (0, 0). [4]

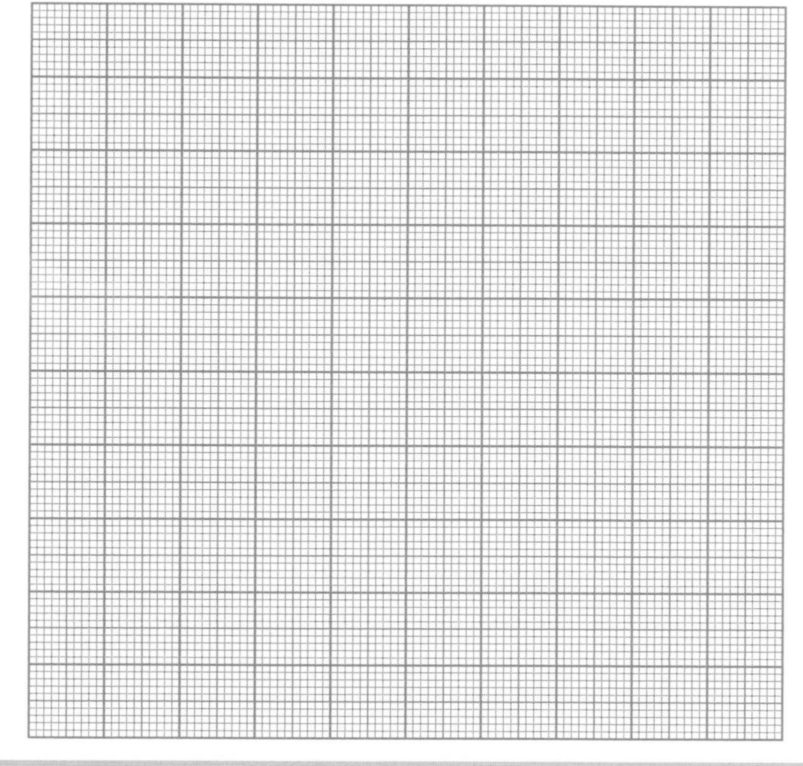

c Determine the gradient G of the graph. Show clearly on the graph how you obtained the necessary information.

G = ... [1]

d Calculate the weight W of the metre rule using the equation

$W = G \times P$

where $P = 1.0\,\text{N}$.

W = ... [1]

e Suggest **one** practical reason why it is difficult to obtain accurate readings for a and b in this type of experiment.

....

.... [1]

f Use the balance provided to measure the mass of the metre rule.

mass = ... [1]

[Total: 11]

Cambridge IGCSE Physics 0625 Paper 51 Q1 May/June 2019

2 In this experiment, you will determine the resistance of a resistance wire.

Carry out the following instructions, referring to Figure 2.

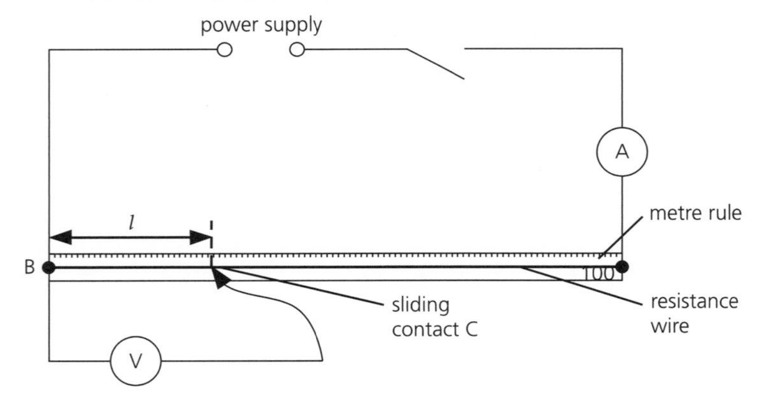

Figure 2

a i Switch on.
Measure the current I in the circuit.

I = ... [1]

ii Place the sliding contact C at a distance $l = 20.0\,\text{cm}$ from B.
Measure, and record in Table 2, the potential difference V across the length l of the resistance wire.

Calculate, and record in Table 2, $\dfrac{V}{l}$.

Repeat the procedure using l values of 40.0 cm, 60.0 cm, 80.0 cm and 100.0 cm.
Switch off. [4]

Table 2

l/cm	V/V	$\frac{V}{l}$/$\frac{\text{V}}{\text{cm}}$
20.0		
40.0		
60.0		
80.0		
100.0		

b Look carefully at the values of $\frac{V}{l}$ in Table 2.

i Tick the box to show your conclusion from the results.

☐ $\frac{V}{l}$ is approximately constant.

☐ $\frac{V}{l}$ is decreasing as V increases.

☐ $\frac{V}{l}$ is increasing as V increases.

☐ There is no simple pattern for $\frac{V}{l}$ in the results. [1]

ii Justify your conclusion by reference to your results.

....

.... [1]

c Calculate the resistance of 100 cm of the resistance wire using the equation

$$R = \frac{V}{I},$$

where V is the potential difference across 100 cm of the resistance wire. Use the value of current I from part **a i**. Give your answer to a suitable number of significant figures for this experiment and include the unit.

R = ... [3]

d In this type of experiment, it is sensible to keep the temperature of the resistance wire as close to room temperature as possible. Suggest **one** way to minimise the rise in temperature of the resistance wire.

....

.... [1]

[Total: 11]

Cambridge IGCSE Physics 0625 Paper 51 Q2 May/June 2019

3 In this experiment, you will determine the refractive index n of the material of a transparent block.

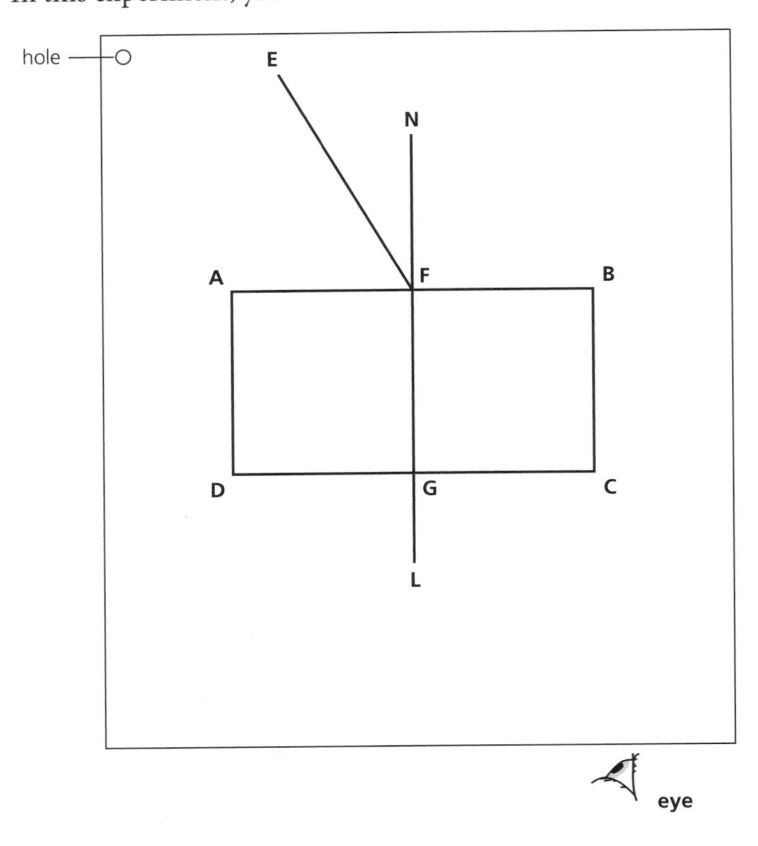

Figure 3

a Carry out the following instructions on a sheet of plain A4 paper. Refer to Figure 3 for guidance.
- Place the transparent block, largest face down, on the sheet of plain paper. The block should be approximately in the middle of the paper.
- Draw and label the outline of the block **ABCD**.
- Remove the block and draw the normal **NL** at the centre of side **AB**.
- Continue the normal so that it passes through side **CD** of the block.
- Label the point **F** where **NL** crosses **AB**.
- Label the point **G** where **NL** crosses **CD**. [2]

b Draw the line **EF** at an angle $i = 30°$ to the normal as shown in Figure 3.
- Place the paper on the pin board.
- Place two pins P_1 and P_2 on line **EF** at a suitable distance apart for this experiment.
- Replace the block and look from the position of the eye shown in Figure 3, to observe the images of P_1 and P_2 through side **CD** of the block. Adjust your line of sight until the images of P_1 and P_2 appear one behind the other.
- Place two pins P_3 and P_4 between your eye and the block so that P_3, P_4, and the images of P_1 and P_2, seen through the block, appear one behind the other.
- Label the positions of P_1, P_2, P_3 and P_4.
- Remove the block and the pins.
- Draw a line joining the positions of P_3 and P_4. Continue the line until it meets the normal **NL**.
- Label the point **J** where the line meets the normal **NL**.
- Label the point **H** where the line meets side **CD**. Draw the line **FH**. [1]

c i Measure and record the length a of the line **FH**.

a = ... [1]

ii Measure and record the length b of the line **HJ**.

b = ... [1]

iii Calculate the refractive index n using the equation

$$n = \frac{a}{b}$$

$n =$ .... [2]

d Repeat the procedure in **b** using the angle of incidence $i = 40°$.

i Repeat the measurements in **c** for $i = 40°$.

$a =$ ....

$b =$ .... [1]

ii Calculate the refractive index n using the equation

$$n = \frac{a}{b}$$

$n =$ .... [1]

e A student carries out this experiment with care and expects the two values of refractive index n obtained in this experiment to be equal.
State **two** difficulties with this type of experiment that could account for any difference in the two values of n.

1 ....

....

2 ....

.... [2]

[Total: 11]

Cambridge IGCSE Physics 0625 Paper 52 Q3 October/November 2018

4 In this experiment, you will investigate how the use of a lid or insulation affects the rate of cooling of hot water in a beaker.

Carry out the following instructions, referring to Figure 4.

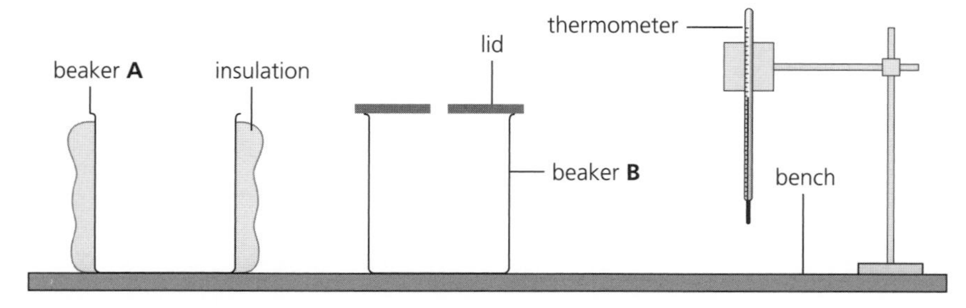

Figure 4

a The thermometer must remain in the clamp throughout the experiment.

- Use the measuring cylinder to pour 100 cm^3 of hot water into beaker **A**.
- Place the thermometer in the water in beaker **A**.
- In the first row of Table 3, record the temperature θ of the water at time $t = 0$ and immediately start the stopclock.
- Record, in Table 3, the temperature θ of the water at times $t = 30$ s, 60 s, 90 s, 120 s, 150 s and 180 s.
- Remove the thermometer from the beaker. [1]

b **i** Repeat **a** for beaker **B**. Ensure that the lid is removed before pouring the hot water into the beaker. Replace the lid immediately after pouring. [2]

ii Complete the headings and the time column in the table. [2]

Table 3

	beaker **A** with insulation only	beaker **B** with a lid only
t/	θ/	θ/
0		

c Write a conclusion stating whether the insulation or the lid is more effective in reducing the cooling rate of the water in the beakers in this experiment.
Justify your answer by reference to your results.

....

....

....

.... [2]

d A student thinks that the experiment does not show how effective insulation is on its own or how effective a lid is on its own.
Suggest an additional experiment which could be used to show how effective a lid or insulation is on its own.
Explain how the additional results could be used.
You are **not** required to carry out this experiment.

additional experiment ....

....

....

explanation ....

....

.... [2]

e Students in another school are carrying out this experiment using equipment which is identical to yours.
State whether it is important for the students to make the initial temperature of the water the same as yours if they are to obtain average cooling rates that are the same as yours. Assume that the room temperature is the same in each case.
Use values from your results for beaker **A** in Table 3 to justify if this factor should be controlled.

statement ....

....

explanation ....

....

.... [2]

[Total: 11]

Cambridge IGCSE Physics 0625 Paper 52 Q2 February/March 2019

5 A student investigates the time taken for metal balls to stop moving after being released on a curved track. Figure 5 shows the shape of the track. The track is flexible, so the shape of the curve can be changed.

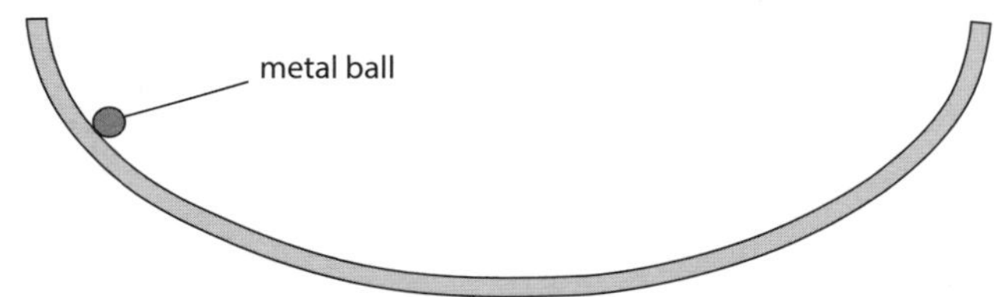

Figure 5

The following apparatus is available:
a selection of metal balls of different masses
the flexible track
clamps to hold the track
a stopwatch
a tape measure
a metre rule

The student can also use other apparatus and materials that are usually available in a school laboratory.

Plan an experiment to investigate a factor that affects the time taken for metal balls to stop moving after being released on a curved track. You are **not** required to carry out this investigation.

In your plan, you should:
- describe how you would expect the balls to move
- explain how you would carry out the investigation
- state which variables you would keep constant and which variable you would change
- draw a table, or tables, with column headings, to show how you would display your readings (you are **not** required to enter any readings in the table)
- explain how you would use your readings to reach a conclusion.

You may add to the diagram in Figure 5 if it helps your explanation.

....

....

....

....

....

....

....

....

....

....

....

....

....

....

....

....

....

....

....[7]

[Total: 7]

Cambridge IGCSE Physics 0625 Paper 51 Q4 October/November 2019

6 A student is investigating whether the type of container affects the time taken for water to be heated from room temperature to boiling point.

The following apparatus is available:

250 cm^3 copper can	measuring cylinder
250 cm^3 aluminium can	thermometer
250 cm^3 glass beaker	tripod and gauze
Bunsen burner	stopwatch

Other apparatus normally available in the school laboratory is also available.

Plan an experiment to investigate whether the type of container affects the time taken for water to be heated from room temperature to boiling point. You are **not** required to carry out this investigation.

You should:

- explain briefly how you would carry out the investigation
- state the key variables that you would control
- draw a table, or tables, with column headings to show how you would display your readings (you are **not** required to enter any readings in the table)
- explain briefly how you would use your readings to reach a conclusion.

....

....

....

....

....

....

....

....

....

....

....

....

....

....

....

....

....

....

.... [7]

[Total: 7]

Cambridge IGCSE Physics 0625 Paper 52 Q4 October/November 2018

Alternative to Practical past paper questions

1 A student is determining the spring constant k of a spring.

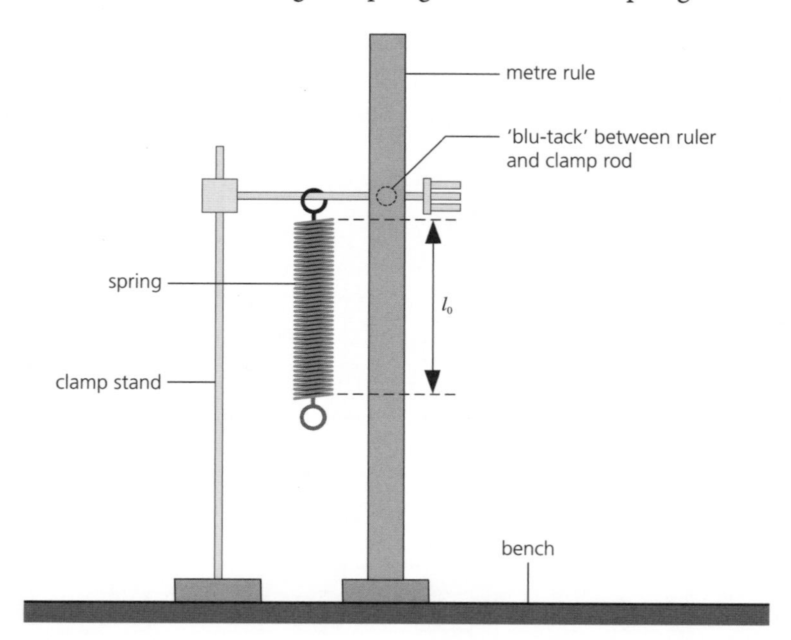

Figure 6

Figure 6 shows the apparatus used.

a On Figure 6, measure the unstretched length l_0 of the coiled part of the spring, in mm. Record this value of length l in Table 4 for $L = 0.00$ N. [1]

b On Figure 6, show how a set-square could be used to take readings in order to determine the length l_0 of the coiled part of the spring. [1]

c The student places a 0.20 N load on the spring. He records the new length l of the spring in Table 4. He repeats the procedure using loads of 0.40 N, 0.60 N, 0.80 N and 1.00 N. All the readings are recorded in Table 4.

i Calculate the extension e of the spring for each value of load L, using the equation $e = (l - l_0)$.
Record the values of e in Table 4. [1]

ii Complete the column headings in Table 4. [1]

Table 4

L/	l/	e/
0.00		0
0.20	31	
0.40	40	
0.60	46	
0.80	55	
1.00	63	

d Plot a graph of L/N (y-axis) against e/mm (x-axis). [4]

e Determine the gradient G of the graph. Show clearly on the graph how you obtained the necessary information.

G = ... [2]

f The gradient G is numerically equal to the spring constant k.
Write down a value for k to a suitable number of significant figures for this experiment.

k = ... N/mm [1]

[Total: 11]

Cambridge IGCSE Physics 0625 Paper 62 Q1 October/November 2018

2 A student is determining the focal length f of a lens.

Figure 7 shows the apparatus used.

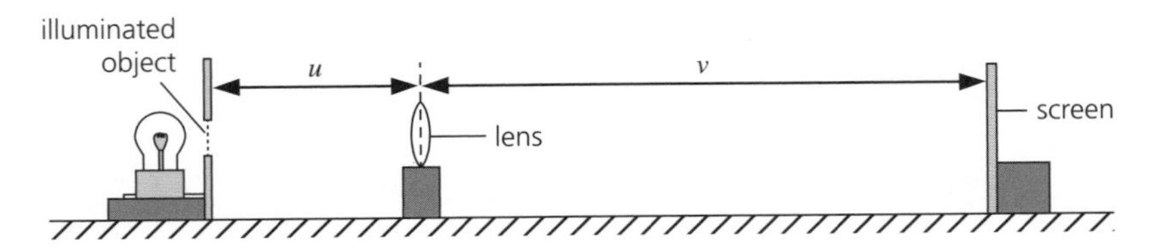

Figure 7

a
- The student places the screen about 100 cm from the illuminated object.
- She places the lens between the object and the screen so that the centre of the lens is at a distance u = 20.0 cm from the object.
- She adjusts the position of the screen until a clearly focused image is formed on the screen.
- She measures the distance v between the centre of the lens and the screen.
- She repeats the procedure using values for u of 22.0 cm, 25.0 cm, 30.0 cm and 35.0 cm.
- The readings are shown in Table 5.

Table 5

u/cm	v/cm
20.0	60.0
22.0	47.1
25.0	37.5
30.0	29.8
35.0	26.3

Plot a graph of v/cm (y-axis) against u/cm (x-axis). You do not need to start your axes at the origin (0, 0). Draw the best-fit curve. [4]

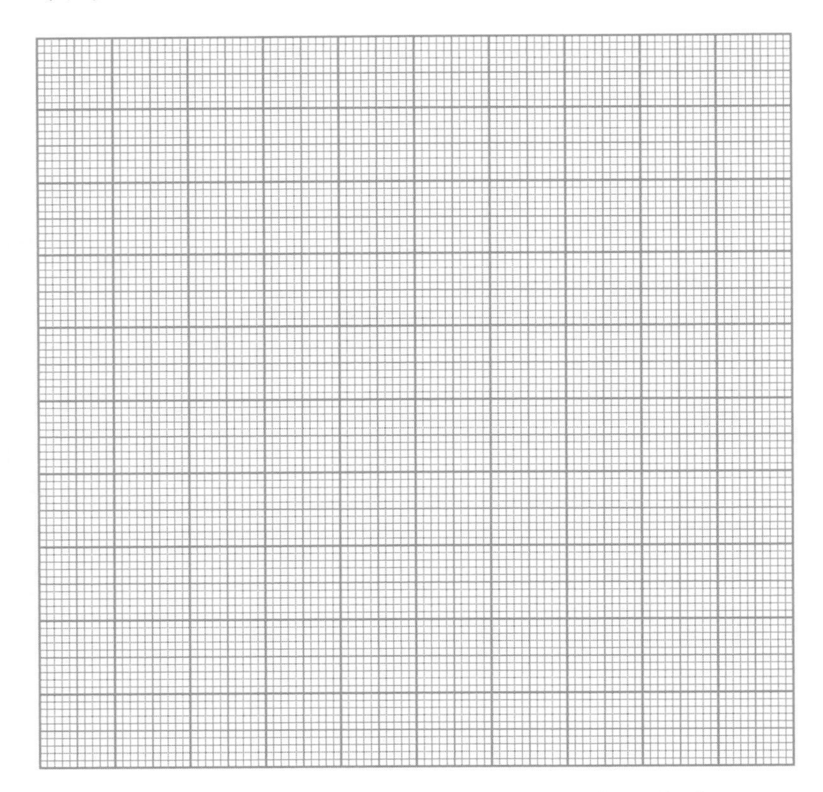

b i
- Mark, with a cross, the point on the graph grid where u = 25.0 cm and v = 25.0 cm.
- Mark, with a cross, the point on the graph grid where u = 35.0 cm and v = 35.0 cm.
- Join these two points with a straight line. [1]

ii – Record u_1, the value of u at the point where the straight line crosses your graph line.

u_1 = ...cm

– Record v_1, the value of v at the point where the straight line crosses your graph line.

v_1 = ...cm [1]

iii Calculate the focal length f of the lens using the equation

$$f = \frac{(u_1 + v_1)}{4}.$$

f = ... cm [2]

c Suggest **two** differences that you would expect to see between the appearance of the illuminated object and the image on the screen.

1 ....

2 .... [2]

d Suggest **two** precautions that you would take in order to obtain reliable readings in this experiment.

1 ....

2 .... [2]

[Total: 12]

Cambridge IGCSE Physics 0625 Paper 62 Q3 October/November 2017

3 Some students are investigating how the volume of water affects the rate at which water in a beaker cools.

They are using the apparatus shown in Figure 8.

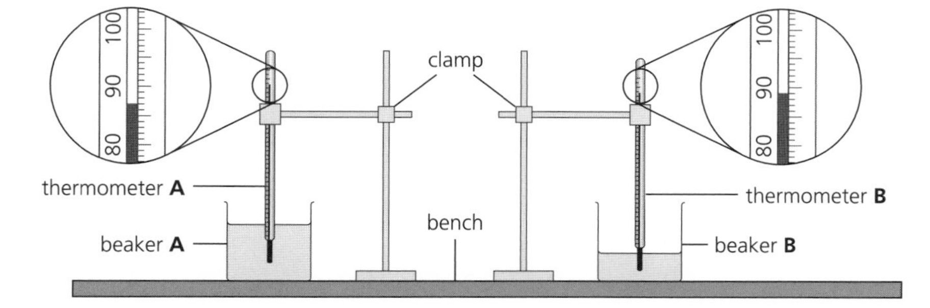

Figure 8

a i 200 cm^3 of hot water is poured into beaker **A** and the initial temperature rises to the value shown on thermometer **A** in Figure 8.
In the first row of Table 6, record this temperature θ_A for time $t = 0$.
100 cm^3 of hot water is poured into beaker **B**. The temperature rises to the value shown on thermometer **B** in Figure 8.
In the first row of the table, record this temperature θ_B for time $t = 0$. [1]

ii The temperatures θ_A and θ_B of the water in each experiment at times t = 30 s, 60 s, 90 s, 120 s, 150 s and 180 s are shown in the table.
Complete the headings and the time column in the table. [2]

Table 6

	beaker **A** with 200 cm³ of water	beaker **B** with 100 cm³ of water
t/	θ_A/	θ_B/
0		
	85.0	86.0
	83.0	83.0
	81.5	80.5
	80.0	78.0
	78.5	76.0
	77.5	74.5

b Describe one precaution which should be taken to ensure that the temperature readings in the experiment are as accurate as possible.

....

.... [1]

c Write a conclusion stating how the volume of water in the beaker affects the rate of cooling of the water. Justify your answer by reference to the results.

....

....

....

....

.... [2]

d i Using the results for 100 cm³ of water, calculate the average rate of cooling x_1 for the **first** 90 s of the experiment. Use the readings from the table and the equation

$$x_1 = \frac{\theta_0 - \theta_{90}}{t},$$

where t = 90 s and θ_0 and θ_{90} are the temperatures at time 0 and at time 90 s.
Include the unit for the rate of cooling.

x_1 = ... [1]

ii Using the results for 100 cm³ of water, calculate the average rate of cooling x_2 in the **last** 90 s of the experiment. Use the readings from the table and the equation

$$x_2 = \frac{\theta_{90} - \theta_{180}}{t},$$

where t = 90 s and θ_{90} and θ_{180} are the temperatures at time 90 s and at time 180 s.
Include the unit for the rate of cooling.

x_2 = ... [1]

e A student suggests that it is important that the experiments with the two volumes of water should have the same starting temperatures.
State whether your values for x_1 and x_2 support this suggestion. Justify your statement with reference to your results.

statement ....

justification ....

....

.... [1]

f Another student plans to investigate whether more thermal energy is lost from the water surface than from the sides of the beakers.
Describe an experiment that could be done to investigate this.
You may draw a diagram to help your description.

....

....

.... [2]

[Total: 11]

Cambridge IGCSE Physics 0625 Paper 62 Q2 February/March 2018

4 A student is investigating electrical resistance.

She uses the circuit shown in Figure 9.

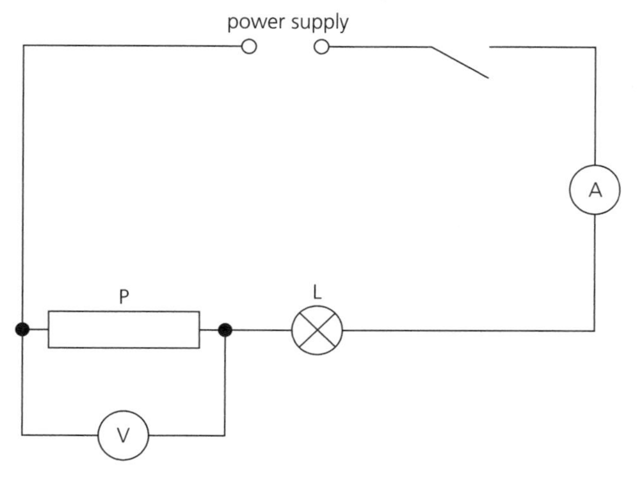

Figure 9

a Write down the readings shown on the meters in Figures 10 and 11.

V_1 = ...

I_1 = ... [2]

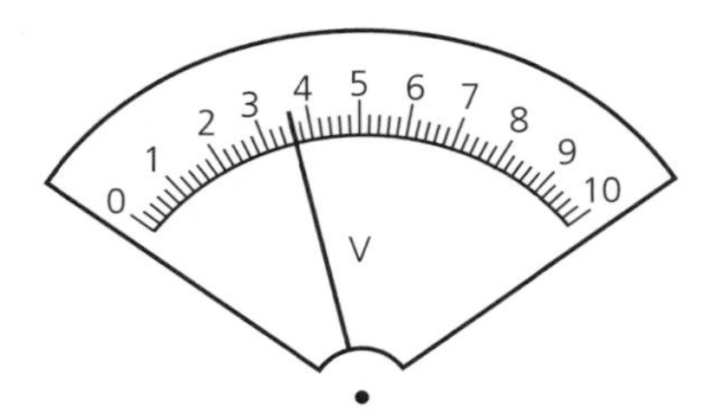

Figure 10

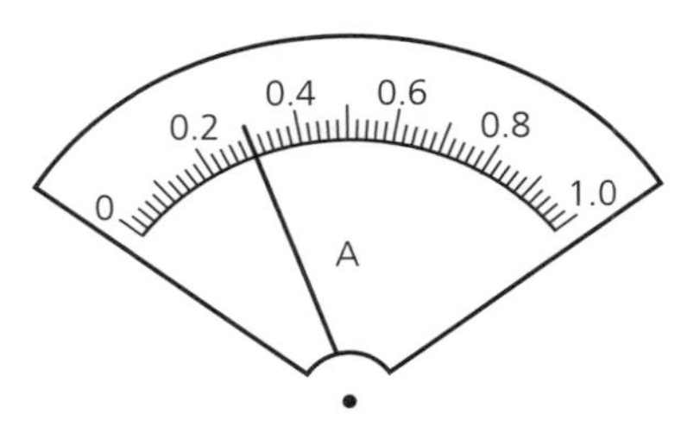

Figure 11

b Calculate the resistance R_1 of the resistor P using the equation

$$R_1 = \frac{V_1}{I_1}.$$

R_1 = ... [1]

c The student connects the voltmeter across the lamp L.
She records the potential difference V_2 across the lamp L.

V_2 = ...2.4 V..........

Calculate the resistance R_2 of the lamp L using the equation

$$R_2 = \frac{V_2}{I_1}.$$

R_2 = ... [1]

d The student replaces the resistor P with the resistor Q.
She records the potential difference V_3 across the resistor Q and the current I_2 in the circuit.

V_3 = ...3.5 V..........

I_2 = ...0.31 A..........

i Calculate the resistance R_3 of the resistor Q using the equation

$$R_3 = \frac{V_3}{I_2}.$$

R_3 = ... [1]

ii State whether the results R_1 and R_3 suggest that resistor P and resistor Q have the same value of resistance, within the limits of experimental accuracy. Justify your statement by reference to your results.

statement ....

justification ....

....

.... [2]

e The student connects the voltmeter across the lamp L.
She records the potential difference V_4 across the lamp L.

V_4 = ...2.5 V..................

She calculates the resistance R_4 of the lamp L.

R_4 = ...8.1 Ω..................

She suggests that the change in resistance of the lamp from part **c** is due to a change in temperature of the lamp filament. Suggest an observation that she could make to confirm that the temperature of the lamp filament changes.

.... [1]

f Complete the circuit diagram in Figure 12 to show that:
- the two resistors and the lamp are all connected in parallel
- the voltmeter is connected to measure the potential difference across the resistors and the lamp. [2]

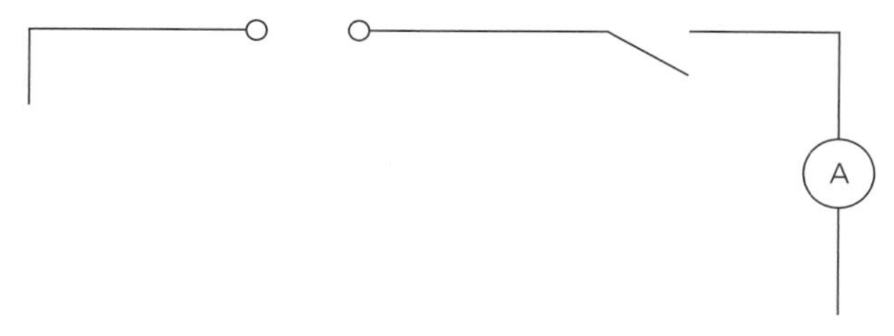

Figure 12

g State the name of the circuit component that you would add to the circuit you have drawn to control the current in the circuit.

.... [1]

[Total: 11]

Cambridge IGCSE Physics 0625 Paper 62 Q3 May/June 2019

5 A student investigates the time taken for ice cubes in a container to melt using different insulating materials on the container.

The following apparatus is available:
a copper container
a variety of insulating materials that can be wrapped round the copper container
a thermometer
a stopwatch
a supply of ice cubes

The student can also use other apparatus and materials that are usually available in a school laboratory.

Plan an experiment to investigate the time taken for ice cubes to melt using different insulating materials.

In your plan, you should:

- draw a diagram of the apparatus used
- explain briefly how you would carry out the investigation
- state the key variables that you would control
- draw a table, or tables, with column headings, to show how you would display your readings (you are **not** required to enter any readings in the table)
- explain how you would use your readings to reach a conclusion.

....

....

....

....

....

....

....

....

....

....

....

....

....

....

....

....

....

.... [7]

[Total: 7]

Cambridge IGCSE Physics 0625 Paper 62 Q4 October/November 2019